'The intersection between geography education, geographies of education, and geographies of children and young people is a particularly vibrant area of scholarship. Yet few publications bring geographers and geography educators together to considers the complex and multi-scalar relationships between children, education and geography, that is until now. Each chapter in this edited volume offers rich insights into research and practice, drawing on a range of conceptual, ideological and methodological traditions to explore the ways in which engagement across these fields can be mutually enriching. The editors describe how the book encourages geography educators and students with an interest in geography, education and young people to engage with questions about the relationships between people, place and nature; the purposes and practices of schooling; and the nature of childhood. Children, education, and geography: Rethinking intersections blends empirical research, theory and practice to stimulate debate on a range of contemporary topics such as decolonising the curriculum, teaching migration and environmental sustainability education. As such it will also be essential reading for new and practicing teachers who want to understand, support and empower the children and young people that they teach.'

Emma Rawlings Smith, Departmental Lecturer in Geography Education, Department of Education, University of Oxford, UK

'In light of growing socio-spatial inequalities and uncertain environmental futures, 'thinking geographically' must mean more than excelling in one specific discipline. Thinking geographically is a skill necessary for students, teachers and scholars of all ages, and this (com)passionately written collection will support teaching, learning and scholarship characterised by thinking geographically together in ways that are at once sensitive to context and expansive across times and spaces.

A key contribution of the collection is the editors' and authors' persistent focus on drawing attention to children and young people's everyday geographies into educational spaces and practices, following the aim set out in the introduction to elucidate and unpack the intersections between 'geography as a discipline and the geographies of everyday lives'. Many of the contributing authors offer practical case examples of integrating insights from diverse child and youth geographies into educational spaces. Perhaps even more useful, however, are the ways that authors draw crucial attention to work not yet done in this regard. The result is a collection that raises urgent and sometimes discomforting questions such as how can educators respond to the climate crisis in hopeful and inclusive ways and how can educators in the global North acknowledge the legacies of epistemic colonialism in inviting learners to participate in building progressive futures.

As well as being of great value to educators, the collection offers rich insights to academic geographers. Specifically, though not limited to the following two areas, the book serves as a bridge between the geographies of education and geographies of children and young people and as such has the potential to generate dialogue and cross-disciplinary working.

Ultimately, this is a book that points ahead, offering suggestions and raising critical questions that have the potential to lead to progressive, inclusive and compassionate ways of knowing, teaching and engaging with the urgent challenges of the day by thinking geographically.'

Catherine Walker, Sustainable Consumption Institute,
University of Manchester, UK

'Highly recommended. A timely book. I learned a lot, thought deeply (again and again), and definitely re-thought my own practice. From the first lines, the affective nature of education is suggested as an essential component of an education if we are to consider a child's full development. This book is shot-through with debates and dialogue with the classroom. Education in this book is seen to be about relationships, agency and nurturing. As Clarke and Witt write so powerfully, we should consider being 'with' our geography 'not in opposition', it is about relationships building over time and 'new ways of being, doing and knowing'. In particular, chapters by Morgan, Lambert & León and Catling & Pike stand out as being useful. Puttick, Chandrachud, Chopra, Robson, Singh and Talks write brilliantly and suggest a conclusion I have come to as well: that the complexity of climate change knowledge means it is unreasonable to place the responsibility on teachers alone as institutions 'have inherent weaknesses for the task of developing local, place-specific and place-relevant information about climate change'. Addressing this should be met by relationships between universities and school teachers, helping show the importance of the varying (and variable) narratives on climate change.'

Anthony Barlow, Principal Lecturer in Early Years and Primary
Geography Education, University of Roehampton, UK

CHILDREN, EDUCATION AND GEOGRAPHY

This book examines the intersections between children, education and geography. With a particular focus on children's geographies and geographies of education, the book draws upon cutting-edge research to consider how geographical education can be enhanced through increased engagement with these fields.

The book is underpinned by the position that the lives of children and young people are inherently geographical, as are educational institutions, systems and processes. The volume explores the ways in which the diverse relationships between children, education and geography can enrich research and work with, and for, children and young people. Chapters in this book consider how in/justices are (re)produced through education. Chapters also explore how insights generated by thinking in, and across, geography and education can be used to support and empower young people in both formal education and in their everyday lives.

Ultimately, this book is written for children and young people. Not as the readership, but as people, often marginalised in decision making at a variety of scales in education, and who, we contend should be at the heart of all educational thinking. The book is of value to undergraduate and post graduate students interested in geography education and children's geographies, as well as teachers of geography, both new and experienced.

Lauren Hammond is Lecturer in Teacher Education at Moray House School of Education and Sport, University of Edinburgh.

Mary Biddulph is former Senior Lecturer in Geography Education at the University of Nottingham.

Simon Catling was a teacher for 13 years in inner London primary schools and moved to Oxford Brookes University in 1984, where he taught geography education and education modules with prospective teachers and education studies students until his retirement in 2012.

John H. McKendrick co-directs the Scottish Poverty and Inequality Research Unit at Glasgow Caledonian University, where he is Professor of Social Justice.

CHILDREN, EDUCATION AND GEOGRAPHY

Rethinking Intersections

Edited by Lauren Hammond, Mary Biddulph, Simon Catling and John H. McKendrick

Routledge
Taylor & Francis Group

LONDON AND NEW YORK

Cover image: © Getty Images

First published 2023
by Routledge
4 Park Square, Milton Park, Abingdon, Oxon OX14 4RN

and by Routledge
605 Third Avenue, New York, NY 10158

Routledge is an imprint of the Taylor & Francis Group, an informa business

British Library Cataloguing-in-Publication Data
A catalogue record for this book is available from the British Library

ISBN: 978-1-032-14746-8 (hbk)
ISBN: 978-1-032-16432-8 (pbk)
ISBN: 978-1-003-24853-8 (ebk)

DOI: 10.4324/9781003248538

Typeset in Bembo
by Apex CoVantage, LLC

To those whose education has not served them well enough and those who work towards a better tomorrow with, and for, children and young people.

CONTENTS

ILLUSTRATIONS

Tables

Figures

CONTRIBUTORS

Tine Béneker is Professor of Geography and Education at the Department of Human Geography and Planning at Utrecht University. Her interests are in future and global perspectives of geography education and the relationships between (disciplinary) academic knowledge, school subject knowledge, and young people's everyday 'knowledges'.

Mary Biddulph is former Senior Lecturer in Geography Education at the University of Nottingham. Mary taught geography in an inner-city school in Derby, before she went on to lead the PGCE Geography course, and whilst at Nottingham, she taught undergraduate and masters programmes. Between 2006 and 2011, with Dr Roger Firth, Mary co-led the Young People's Geographies project (funded by the Action Plan for Geography), the focus of which was recontextualising academic geography research into the lives of young people for the school geography curriculum. Mary was Editor of the journal *Teaching Geography* and was President of the Geographical Association in 2017.

Rachel Brooks is Professor of Sociology and Associate Dean for Research and Innovation at the University of Surrey. She is also Editor-in-Chief of *Sociology* and Co-editor of the Routledge/SRHE 'Research into Higher Education' book series. She has published widely in the sociology of higher education: recent books include *Student Migrants and Contemporary Educational Mobilities* (2021, with Johanna Waters) and *Reimagining the Higher Education Student* (2021, with Sarah O'Shea).

Simon Catling Following graduation in geography and education and 13 years of teaching in inner London primary schools, Simon Catling moved to Oxford Brookes University in 1984, where he taught geography education and education modules with prospective teachers and education studies students until his retirement in 2012. He also led masters courses on curriculum, teaching, and learning. From 1995, he worked in senior management in the School of Education, as Deputy and Head of Department and as Associate Dean of Education. He was awarded Professorship in 2001, and an Emeritus Professorship by the university in 2012, for his work, scholarship, and research, largely contributing to primary geography curriculum, teaching, learning and resources, including materials for primary children, for teachers, and teacher educators and researchers. President of the Geographical Association in 1992–1993, he was awarded an honorary membership in 2017, for his work nationally and internationally in geography education. He has published many articles for teachers and researchers, books to support the teaching of geography, including *Understanding and Teaching Primary Geography* with Tessa Willy (2018) and *The Everyday Guide to Primary Geography: Locational Knowledge* (2020), and the *Outset Geography* (1982–1985) and *Mapstart* (1985, 2010) texts for primary children.

Paloma Chandrachud As an environmental studies graduate from FLAME University (India), Paloma Chandrachud is keen to explore the intersections between climate science, conservation, and communication. Her areas of interest are understanding the existing knowledge and awareness about climate change, education, and existing policies. She is currently working with Trans-disciplinary Research Oriented Pedagogy for Improving Climate Studies and Understanding (TROP ICSU) led by the International Union of Biological Sciences (IUBS) and FLAME University, India. As part of TROP ICSU, she has been involved in creating teaching resources and data configuration for the online website. As she continues to work in the field of climate education, she hopes to establish herself in the field of communication and conservation.

Rahul Chopra is Director of the Centre for Sustainability, Environment and Climate Change and Associate Professor of Environmental Studies at FLAME University, India. He is also the coordinator and co-lead of a global climate change education project of the International Union of Biological Sciences (IUBS) that aims to integrate climate change education in the curriculum across the world. He is a member of the Expert Group on COVID-19 in Pune as part of the Pune Knowledge Cluster; Ambassador of The Earth Project; Visiting Professor at Ashoka University, and a member of the Climate Collective Pune. He was also the coordinator and convener of the National Resource Center on Climate Change at the Indian Institute of Science Education and Research (IISER), Pune, an initiative of the Ministry of Human Resource Development (MHRD), Government of India.

Rahul Chopra's interests are multidisciplinary and include curriculum development in Earth and Environmental Studies; the use of satellite-derived remotely sensed and in situ data to evaluate our changing environment; historical GIS and the spatial humanities; field-based geological and environmental studies; and the use of high-resolution chemical analyses instruments and data to study various earth and environmental processes. He received his PhD in Geophysical Sciences from the University of Chicago, USA.

Helen Clarke and Sharon Witt are independent scholars who have both worked as teachers in primary school and as academic tutors within Initial Teacher Education. Their work engages with playful, experiential approaches to place responsive learning. They are currently exploring innovative practices that consider the potential of a relational approach to natural encounters through signature 'pedagogies of attention'. Their research interests include geography, science, transdisciplinary and arts-based practices, curriculum making, and teacher professional development. You can follow their work on Twitter @Attention2place.

Ria Dunkley is Senior Lecturer in Geography, Environment and Sustainability within the School of Education at the University of Glasgow. She has been investigating post-humanism and people's relationship with the environment since 2011. Her research examines how climate and environmental crisis is driven by human relationships with the natural world and identifies effective formal and informal pedagogies (known as ecopedagogies) that raise the consciousness of this relationship to strengthen nature connectedness. Since 2016, she has researched the effects of participation in Citizen Science (CS) projects. She has demonstrated that citizen scientists participating in water quality monitoring and phenology projects are motivated to connect to the natural surroundings. This desire is related to natural world relationships rooted in childhood experiences.

Lauren Hammond is Lecturer in Teacher Education at Moray House School of Education and Sport, University of Edinburgh. Prior to this, Lauren worked as a secondary school geography teacher in the UK and Singapore, before working at IOE, UCL's Faculty of Education and Society for eight years. At IOE, Lauren worked in teacher education, convened an undergraduate module for students in UCL's geography department 'geography education', and supervised students at masters and doctoral levels. She was also co-chair of IOEs Early Career Network. Lauren is committed to researching with, and for, young people, and her research straddles the fields of children's geographies, children's rights, geography, and education. Lauren is Senior Fellow of the Higher Education Academy (SFHEA) and Fellow of the Royal Geographical Society (FRGS). She serves as Deputy Secretary for the Royal Geographical Society's Geography and Education Research Group and Membership Office of their Children, Youth and Families Research Group. She is a member of the Geography Education Research Collective (GEReCo).

Grace Healy is Curriculum Director at David Ross Education Trust. In her previous role, she led the geography subject community across a trust of 13 primary and secondary schools and contributed to the leadership of a SCITT. She is currently undertaking a PhD at IOE, UCL's Faculty of Education and Society. Grace is chair of the Geographical Association's Teacher Education Phase Committee and Honorary Secretary (Education) for the Royal Geographical Society (RGS). She is Treasurer for the Geography Education Research Collective (GEReCo) and the RGS's Geography and Education Research Group. Moreover, she serves on British Educational Research Association's Publication Committee and on the editorial boards of *The Curriculum Journal* and the *London Review of Education*.

Peter Hopkins is Professor of Social Geography at Newcastle University and during 2020–2021, he was Distinguished International Professor at Universiti Kebangsaan Malaysia. His research and teaching expertise focuses upon young people, place, and identity; geographies of race and religion; refugee and migration studies; and the intersections between masculinities and ethnicities. He was elected a Fellow of the Academy of Social Sciences in 2018 and was awarded a Newcastle University Vice Chancellor Distinguished Teacher Award in 2011. His previous publications include *Young People, Place, and Identity* (Routledge) and *Social Geographies: an Introduction* (Rowman and Littlefield) with the Newcastle Social Geographies Collective and *Children, Young People, and Critical Geopolitics* (Ashgate) with Matthew Benwell. He recently co-authored (with Simon Tate) *Studying Geography at University: How to Succeed in the First Year of Your New Degree* (Routledge).

Peter Kraftl is Professor of Human Geography at the University of Birmingham, UK. Although a geographer, he is an interdisciplinary scholar of childhood and youth, with a particular interest in the emotions, affect, materialities, and embodied practices that constitute children's lives. He has worked broadly on children's experiences of urban environments and environmental issues – from young people living in newly built, master-planned urban communities in the UK, to children's positioning within the 'food-water-energy nexus' in Brazil. He has published eight books and over 100 journal articles and book chapters. His most recent book, *After Childhood*, was published by Routledge in 2020. He is currently co-leading major funded projects looking at children's role in the future of treescapes in the UK, and at children and public health. Peter has edited the journals *Children's Geographies* and *Area* and is currently Honorary Secretary for the Royal Geographical Society (with IBG) Research and Higher Education Division. Peter is a Fellow of the RGS and the Academy of Social Sciences.

David Lambert is Emeritus Professor of Geography Education at IOE, UCL's Faculty of Education and Society, London. He graduated from the University of Newcastle, completing a PGCE at the University of Cambridge and a PhD at the University of London. He was a secondary school teacher for 12 years, becoming deputy principal of a comprehensive school. He wrote award-winning school textbooks and became a teacher-educator from 1986, since when he has published widely on the curriculum, pedagogy, and assessment of geography in education.

He was Chief Executive of the Geographical Association from 2002 to 2012, being appointed Professor of Geography Education in 2007, before retiring in 2018.

Kelly León teaches grade 9 human geography at a high school in south San Diego County and co-leads the *Generation Global* cohort of teacher candidates at San Diego State University (SDSU), where she teaches credential and MA courses. She completed her undergraduate degree, bilingual teaching credential, and MEd in Policy Studies in Language and Cross-Cultural Education at SDSU. She is currently a doctoral candidate in the Education for Social Justice PhD programme at the University of San Diego. Her dissertation research is focused on collaborative efforts to conceptualise and enact ethnic studies curriculum in secondary schools.

Fran Martin is Honorary Research Fellow in the Graduate School of Education, University of Exeter, UK. She has worked in education for 40 years, first as a primary school teacher and then, since 1993, as a teacher educator in the higher education sector. Fran has a background in geographical and global education and for the last six years her research has focused on critical interculturalism and de/colonial pedagogies.

John H. McKendrick was Founding Member of the Geographies of Children, Youth and Families Research Group (of the RGS-with-IBG) and is currently Ordinary Committee Member. He co-directs the Scottish Poverty and Inequality Research Unit at Glasgow Caledonian University, where he is Professor of Social Justice. Much of John's work involves straddling the academic and policy communities concerned with tackling child poverty and promoting children's play in Scotland. John has a long-standing interest in education in schools, and over recent years, he has been exploring how and why children's geographies are of value to school geography.

David Mitchell is Associate Professor of Geography Education at IOE, UCL's Faculty of Education and Society. David's research interests include teacher agency in the curriculum and how subject teachers develop and use their knowledge in education for sustainability. David was Principal Investigator of the 'GeoCapabilties 3' project (2018–2021). He is the author of *Hyper-Socialised* (2020), a book exploring the potential of teachers as autonomous 'curriculum makers' in late capitalism.

John Morgan taught geography in schools and colleges in London before leading the Secondary Geography PGCE course at the University of Bristol. He is Professor of Geography and Environmental Education the UCL-IOE and Head of the School of Critical Studies in Education at the University of Auckland. His most recent books are *Teaching Secondary Geography as if the Planet Matters* (Routledge, 2012) and *Culture and the Political Economy of Schooling: What's Left for Education?* (Routledge, 2018).

Susan Pike After taking her geography degree and working for a number of years, Susan became a secondary geography teacher in 1994. She worked in a number of schools in Hampshire, including as Head of Geography during the 1990s. Susan moved to Ireland in 2000, and became a lecturer in geography education. Susan has taught on a range of programmes over the years, designing the geography input for the ITE programmes as well as co-creating the masters in geography and history education. She currently teaches geography, environmental and outdoor education on the early childhood and primary teacher education programmes at Dublin City University. She teaches masters programmes on climate change and poverty and social inclusion. Susan's research includes teacher education, teaching and learning geography in schools, sustainability education, as well as children's geographies. Her book *Learning Primary Geography: Ideas and Inspirations From Classrooms* was warmly welcomed in 2016, and she is currently writing a second-level version of this book. She has written resources for schools including the bestselling *Folens/ Phillips Primary Atlas* and activity book *Atlas Hunt*. Susan was President of the Geographical Association in 2020–2021, and she remains an active member working on a number of diversity, inclusion, and curriculum projects.

Fatima Pirbhai-Illich, a transnational feminist, is Professor of Language and Literacy Education in the Faculty of Education at the University of Regina, SK, Canada. She has worked in tertiary-level teacher education for over 25 years and her community-based research focuses on social and human justice. Over the past decade, Fatima has been working towards de/colonial pedagogies in language and literacy education.

Steve Puttick is Associate Professor of Teacher Education at Oxford University's Department of Education and Fellow of St Anne's College. He researches at the intersection between the academic discipline and school subject of geography. Recent research includes the GCRF-funded projects: Climate Change Education Futures in India; Cultural Heritage and Curriculum Making in Kolkata; and Resilient Lagoons in West Africa. Steve serves on the editorial board of the journal *Geography* and is Chair of the Geography Education Research Collective (GEReCo/IGU-CGE).

James Robson is Co-Director of Oxford University's Centre for Skills, Knowledge, and Organisational Performance (SKOPE) and Lecturer in Higher Education at Oxford's Department of Education, where he leads the MSc in Higher Education. His research focuses on the political economy of tertiary education systems with a particular focus on policy, structures, the intersection between education and employment, skills supply and demand, research eco-systems, social justice, and sustainability. His recent projects have focused on skills supply for the green economy; skill needs in big business contexts; global research ecosystems; and widening participation and access. He has received major research funding from the

ESRC, the AHRC, the GCRF, the Edge Foundation, the Royal Society, the British Academy, and the Office for Students and Research England.

Sanjana Singh An avid learner who is a glutton for knowledge, Sanjana Singh constantly upskills to be creative, innovative, collaborative, and empathetic. She graduated as a sustainability researcher and scientist from United Nations University (UNU), Japan, with interests in education, public policy, indigeneity, and corporate social responsibility. She is currently working in Climate Change Education with Trans-disciplinary Research Oriented Pedagogy for Improving Climate Studies and Understanding (TROP ICSU) led by the International Union of Biological Sciences (IUBS). As a part of TROP ICSU, she has been involved in creating over one hundred teaching resources and has held workshops for 'Train the Trainer' module. Her goal is to establish herself in the field of climate communications.

Isobel Talks is a DPhil student in the Learning and New Technologies group, which is part of the Department of Education at the University of Oxford. Her thesis critically explores the 'gender data revolution' in international development through an in-depth case study of a smartphone-based data collection project working with young women in Bangladesh. During her time in Bangladesh, Isobel contributed to teaching and research activities at the Centre for Sustainable Development at the University of Liberal Arts (ULAB) in Dhaka as a 'Visiting Researcher'. These experiences in Bangladesh motivated Isobel to seek out further opportunities to utilise her skills and experience to help fight the climate crisis and support the movement for climate and environmental justice. This led to her working as a research assistant for the Nuffield-funded 'Trust and Climate Change: Information for Teaching in a Digital Age' project and now the GCRF-funded 'Climate Change Education Futures in India' project. Alongside her academic work, Isobel is also an independent consultant and researcher for organisations including Plan International, DFID, and Save the Children.

Simon Tate is Professor of Higher Education at Newcastle University and a National Teaching Fellow. From 2016 to 2020, he was a member of the *Royal Geographical Society's Accreditation Panel for Undergraduate Degrees* and from 2021 to 2022, he was a member of the QAAs Subject Benchmark Panel for Geography. He delivers the Royal Geographical Society's annual New to Teaching Geography workshops, and in 2021, he co-authored the RGS publication *New to Teaching Geography: a Practical Guide for Higher Education Teaching Assistants, Teaching Fellows and Demonstrators*. Before taking up his current appointment, Simon taught geography in schools in the north-east of England, and his current role sees him teaching extensively on the first year of Newcastle University's geography degree programmes. His main research interest is the social and academic transition from school to university geography – particularly how these are experienced by non-traditional students. His latest textbook *Studying Geography at University: How to*

Succeed in the First Year of Your New Degree (co-written with Prof. Peter Hopkins) was published by Routledge in 2020.

Johanna Waters is Professor of Human Geography at University College London. She is Co-Director of the Migration Research Unit and Director of the MSc Global Migration. She has published widely on aspects of transnational migration, families and households, and international and transnational education. Her latest book (with Rachel Brooks, 2021), titled *Student Migrants and Contemporary Educational Mobilities*, was published by Palgrave.

INTRODUCTION

1

THE CHILD AND THEIR (GEOGRAPHICAL) EDUCATION

Lauren Hammond, Mary Biddulph, Simon Catling, and John H. McKendrick

Introduction

> How can a bird that is born for joy,
> Sit in a cage and sing?
>
> (Blake, 1789)

In his poem 'The School-Boy', Blake uses the metaphor of a caged bird to represent a child attending school, forced to leave behind the wonders and pleasures of the summer morning to which he had awoken. Blake contrasts the eighteenth-century rural idyll which the boy calls home with the school as an institution within which the boy is 'caged' and dominated by the 'cruel eye' of his teacher. We begin the book with an extract from this poem – not because we believe that all education is like this – but because it represents the deeply affective nature of educational spaces, teaching, and schooling for children and (young) people and also for their parents/ carers, and those who work with and for children. Written over two centuries ago, the poem encourages us to engage with questions about the relationships between people, place, and nature; the purposes and practices of schooling; and the nature of childhood. These are questions which still resonate today when considering how best to support and empower children and young people through education in their lives and futures. These questions carry an especially heavy weight in the period of multi-species urgency (Haraway, 2016b) and intersecting injustices (Puttick, 2022) which characterise the present.

As the poem highlights, education matters to people's feelings, identities, relationships, and spatialities. Education also matters because of the opportunities, communities, spaces, and places it (re)produces and supports a person to access. Education matters because of the ideas, questions, skills, and knowledges with which a person engages and the futures this makes possible for themselves as an

DOI: 10.4324/9781003248538-2

individual, for society and for the Earth. As the title of this introductory chapter suggests, geography is not only important as a component of education – as a subject with which children and young people engage in schools or universities – it is also significant in helping us to better understand educational spaces, places, systems, processes, and institutions.

Motivated by considering how we can best support and empower children and young people to 'sing' in, and through, their education, this book is a contribution to the growing body of literature that considers the complex and multi-scalar relationships between geography and education (Taylor, 2009; Brock, 2016; Brooks and Waters, 2017; Janhnke et al., 2019; West et al., 2020; Finn et al., 2021; Freytag et al., 2022; Puttick, 2022). We focus specifically on the intersections between children, education, and geography. As the reason why a society educates and why many educators choose their careers, children and young people are central to both educational practice and academic and political debates about education. Yet, how the child is constructed and represented in education varies between places and across time-space, with children's lives and identities often socially and/or spatially shaped by education (Hopkins, 2010; Oswell, 2013). These are important considerations, as the social construction of 'the child' in educational policies and practices has, at times, led to children being subordinated in both education and society, for example, through corporal punishment in classrooms and by gender or 'ability' streaming which predetermines the ideas, knowledges, and/or skills with which a child can engage through education.

Education is infused with 'moral geographies', which are shaped by imagination, ideology, and axes of 'social difference and power' and connect to 'wider ideas of citizenship, belonging, landscape and nationhood' (Mills, 2022: p. 9). These moral geographies impact on the design of educational institutions, structures, and processes, which can ultimately impact on a child's agency, feelings, and engagement with education. However, children and young people are not passive recipients of education, rather they are beings, becomings, and doings, 'active in the construction and the reimagining of their spaces' (Aitken, 2018a: p. 11). Respecting the child in this way means valuing them as person in both education and everyday life and actively considering the relationships between these two spaces to support and empower children and young people in their lives and futures. This, we argue, requires rethinking the intersections between children, education, and geography.

Intersections and relations matter on many levels; from the individual to the societal, from the practical to the conceptual. How we think about relations and how we tell the stories of those relations also matters to their (re)production and more broadly to social practices. As Haraway (2016a) explains when drawing on the work of Marilyn Strathern to explore the relationships between humans and nature in this era of multi-species urgency, 'it matters what matters we use to think other matters with' (p. 12). Haraway's argument is helpful when applied to rethinking the intersections between children, education, and geography. For example, if we tried to examine education without using geography's concepts, ideas, and methods, or without truly engaging with the experiences and imaginations of children and young people, then we are likely to overlook critically important issues such as injustices in the spatial provision of education and their impacts on communities.

We begin this introductory chapter by critically engaging with the present time-space. We then reflect on current literature and practice in, and across, the intersections between children, education, and geography. Following this, we introduce the three sections which frame the book: *Section I: geographies of education and educational spaces; Section II: children's geographies and their significance in, and to, everyday life and education;* and *Section III: progressive geographies in education.* Finally, we offer a statement on ethics and language used in the book.

The importance of rethinking intersections in the present time-space

We begin by rethinking the intersections between children, education, and geography in the present time-space, because how could we not? We live in a period of intersecting crises – COVID-19 and the threat of future pandemics; the climate and ecological emergencies; societies permeated by structural and everyday injustices; and conflict among people and between states – all of which directly impact on lives and futures.

As Latour (2018) explains, the Earth itself (the Terrestrial) can now be seen as a political actor, participating in history and reacting to human actions:

> Formerly, it was possible to say that humans were "on earth" or "in nature," that they found themselves in the "modern period" and that they were "humans" more or less "responsible" for their actions. One could distinguish between "physical" geography and "human" geography as if it were a matter of two layers, one superimposed on the other. But how can we say where we are if the place "on" or "in" which we are located begins to react to our actions, turns against us, encloses us, dominates us, demands something of us and carries us along in its path?
>
> *(p. 41)*

For Latour (2017), the term crisis itself is problematic – crises pass – this is 'profound mutation of our relation to the world' (p. 15). The world has changed, and so must our ways of thinking about and acting in the world. As Haraway urges us we must *stay with the trouble*, we must be truly present, for 'these times are ours' (Haraway, 2016b: p. 40). These crises result in a need to rethink human–nature relations, they require collaboration between people, more-than-human beings, and also between ideas and communities. For 'alone, in our separate kinds of expertise and experience, we know both too much and too little, and so we succumb to either despair or hope, and neither is a sensible attitude' (Haraway, 2016a: p. 4).

It is of critical importance that children and young people are informed, supported, and empowered in these times. Children and young people currently navigate the 'competing contortions and entanglements of climate fact, value, and concern' (Rousell and Cutter-MacKenzie-Knowles, 2020: p. 192), with which they engage through conversations with their friends and families, (social) media, and educational resources and dialogues. Children are often expected to 'grapple with various futures presented to them and what might be done to achieve them' (Walshe and Sund, 2022: p. 110) and may experience emotional responses to these discourses (Ibid.). The lives and futures of many children and young people around the world have also been, or will be, altered by temperature rises and ecological changes on multiple scales; what they eat, how they keep warm/cool, where they live, and their relationships with, and to, nature and environments, will all be altered. Yet, as Yusoff (2018: p. 2) explains 'as the Anthropocene proclaims the language of species life – *anthropos* – through a universalist geologic commons, it neatly erases the histories of racism'. Just as the anthropogenic causes of these crises are not universal, neither will their impacts be, 'for it is those people that are lacking resources who are the principal casualties of slow violence' (Nixon, 2011: p. 4). Put another way, the 'other' here will likely be racialised, gendered, and disproportionally impact on those living in poverty.

Nixon's concept of slow violence is helpful here in considering how these crises emerge and, at times, have seemingly been 'allowed' to continue both above and below the metaphorical radar. Nixon (2011) defines slow violence as:

> Violence that occurs gradually and out of sight, a violence of delayed destruction that is dispersed across time and space, an attritional violence that is typically not viewed as violence at all.
>
> *(p. 2)*

Examples of slow violence include the exposure of children and young people to air pollution when living on busy roads and their everyday entanglements with toxic substances like plastic (Kraftl, 2021). Exposure to air pollution over a sustained period can impair health, with mortal consequences in more extreme cases. These are 'the long dyings – the staggered and staggeringly discounted casualties, both human and ecological' (Nixon, 2011: p. 2). Slow violence may be hard to immediately see, measure, and respond to; and it may be deliberately or inadvertently ignored or discounted. Drawing on the work of Nixon (2011) and Haraway (2016a), Kraftl (2021) asks us how we can learn to live with these changes to the world and how we can *stay with the trouble*.

Here, we contend that geography education is of the utmost importance in engaging children and young people in, and with, a critical conversation about the Earth and the future. Both geography as a discipline and the geographies of everyday life, and significantly the relationships between them are central to these discussions. Put another way, understanding children's geographies and geographies of injustice (including those related to education and the intersecting crises discussed earlier) can help inform educators in their practice. In turn, geography education can inform and empower children in their everyday lives and futures. This is because geography as a discipline can help us to understand, explain, predict, and mitigate these intersecting crises. Geography education in schools has an important role in telling these stories and supporting children to use disciplinary thought to investigate and think about the world, their actions in it, and the futures they want to co-create for themselves, society, and the Earth. This includes active consideration of not only what is taught but also how. For example, as Dunkley (2022) details, educational, mental health, and wider social benefits are accrued when children are supported to engage with nature and knowledge production processes through citizen science. Strategies such as arts-based pedagogies and co-production-based research are also highlighted as being beneficial in supporting children and young people to make meaning, discuss their feelings, and/or to empower them as actors in education, research, and the world (Rousell and Cutter-MacKenzie-Knowles, 2020; Rushton et al., 2021; Walshe, 2017).

However, neither the place of geography (Lambert and León, 2022) nor teaching about these areas through education and sustainability education in schools is secure. For example, in England, education at all levels from primary to higher education has been increasingly marketised. This process has occurred as the state continues to push for greater influence in, and control over, teacher education (DfE, 2021) and even political impartiality in schools (DfE, 2022). Whilst some have argued against transgressing the 'line' between education and politics – particularly related to activism in education (Standish, 2021) – we argue that education is inherently political, and it is necessary to *stay with the trouble*. As a state-funded public service, schools are institutions which formalise ideas about what it means to be a 'good' child or citizen (Mills, 2022). Education plays a significant role in social and political (re)production in society (Giroux, 1982), and 'by appearing to be an impartial and neutral "transmitter" of the benefits of a valued culture, schools are able to promote inequality in the name of fairness and objectivity' (p. 97). Thus, both Standish's (2021) plea for neutrality, and more broadly in his othering and grouping of what he terms 'curriculum decolonisers' (p. 142), he is effectively ignoring 'the conflict between the powerful and the powerless' in terms of debates about the ecological and climate emergencies, and the relationships between geography, education, and in/justice, which ultimately 'means to side with the powerful, not to be neutral' (Freire, 1984: p. 112).

Here, it is important to recognise that children and young people are not passive, they have 'unique perspectives and political agencies' (Skovdal and Benwell, 2021: p. 259). From discussions with their families to climate school-strikes, children contribute to debates and actions in the world (Catling and Pike, 2022), and in doing so they shape spaces and ideas at multiple scales. Children and young people can, and do, contribute to a better tomorrow for themselves, society, and the Earth. Yet, everyday and structural injustices – including those related to class, 'race', gender, dis/ability, and sexuality – which are socially and spatially (re)produced in both the contestations of daily life and education, alongside distrust of political systems can impact on children's agency and activism (Walker, 2021). Socio-spatial and political structures also impact on access to facilities and opportunities, varying from access to schooling itself to the type of resources children engage with in education.

Rethinking the intersections between children, education, and geography can help us better understand structural and everyday injustices and how they are (re)produced in, and through, education. Rethinking these intersections can also help us to imagine progressive futures for (geography) education in which injustices are challenged, and in which children and young people are supported and empowered as actors in both education and everyday life. For some children (and their families and teachers) this can be discomforting. Some view that children should be protected from such engagement, but this means protecting them from learning that there are challenging and contentious dimensions to everyone's lives, which in itself is not liberating and is problematic educationally. There are sensitivities to treat carefully when exploring injustices, but these can be tackled with even the youngest children (Kavanagh et al., 2021; Dolan, 2022). In the next section, we examine the work done so far to rethink these intersections.

Children, education, and geography: it matters what matters we think other matters with

Just as we might learn about geography through education in lessons in schools or lectures at university, because education is a fundamental part of most societies, it 'has something to say about how the world works – its human geographies' (Brooks and Waters, 2017: p. 4). Over recent years, there has been increasing research interest into the relationships between geography and education. Significantly, both geography and education are exemplars of what Brock (2016: p. 10) terms *composite and integrative disciplines*, with their identities resting 'on a particular array of contributing subjects and disciplines'. Brock is clear to point out that this does not deprive either discipline of a 'distinctive character or essence' (Ibid.) and significantly that education is both a discipline and phenomena (i.e., actual teaching and learning). This means that the (potential) relationships between geography and education are complex and multi-scalar, varying from examinations of lived experiences and micro-geographies of schooling, to macro-analyses of the globalisation of higher education (Taylor, 2009).

In the discipline of geography, the fields that most directly engage with education are *children's geographies* (Horton et al., 2008; Aitken, 2018b) and *geographies of education* (Pini et al., 2017; Waters, 2018), both of which have developed since the 1960s. Both children's geographies and geographies of education are rich and diverse methodologically and substantively, and research in these fields can help us to better understand education and its relationships with, and to, everyday life and society. As Kraftl et al. (2021: p. 15–16) explain when considering the evolution of research in geographies of education:

> Geographers have made a distinct contribution to studying the spatialities of education through key geographical tropes such as space, place, and

scale: from a focus on spatial science and quantitative approaches to mapping school access or segregation, to an examination of identities and processes played out in education spaces.

Equally, research in the subdiscipline of children's geographies has the power and potential to enable us to better understand children's experiences and imaginations of education from their own perspectives. The value of this ultimately lies in having 'concern for education's future impacts, encouraging us to engage with young people as knowledgeable actors whose current and future life worlds are worthy of investigation' (Holloway et al., 2010: p. 294).

Education is also a fundamental part of how geography as a discipline is reproduced as students are inducted into its ideas, methods, and ways of thinking through educational programmes, curricula, teaching, and assessment. As such, there is both research and practical interest into how best to teach students geography in universities, including through the field of *pedagogic research* (Finn et al., 2021). Geographers working in other subdisciplines may also be keen to share their research with teachers and young people, and to inform and support practice in schools and student transition to university (Tate and Hopkins, 2019).

In the discipline of education, geography is most often considered as part of the field of *geography education*. Geography education expanded as a field of research and teaching in the UK after 1945 (Butt, 2019), with research in the field often focusing on school geography (how geography is constructed, represented, taught, learnt, and assessed in schools) and/or (initial) teacher education. One significant area of debate in the field which focuses on the intersections between geography and education is active consideration of causes, impacts, and nature of 'the gap' (Butt, 2019) between geography as an academic discipline and school subject. Here, research has also examined if and how the school subject connects to and represents children's everyday geographies and considered how the construction of 'the child' in schooling is often divergent from disciplinary debate in geography (Catling and Martin, 2011), in which the child is recognised and celebrated as being, becoming, and doing (Aitken, 2018a).

Research in geography education is often conducted by those working in teacher education in universities, and increasingly also by those working in schools and other educational settings. Research in the field has an important role to play in better understanding the nature of school geography, supporting and empowering teachers in their practice, informing policy and debate, and advocating for change in geography education (Lambert, 2010), which includes recognising children as geographers (Catling, 1988; Catling and Willy, 2018). However, geography education as a field of research has been described as relatively small scale, regularly self- or un-funded, and piecemeal (Lambert, 2010; Butt, 2019). The educational and market agendas of the state, performativity regimes (for example, Ofsted in the English Context), and neoliberalism also directly influence how school geography is constructed and experienced, and who can study and teach geography. Significantly, beyond research and teaching in geography education, geography is not consistently recognised in disciplinary discussions or teaching modules in faculties of education in the same way as sociology, psychology, or history (Brock, 2016). This, we argue, is a significant omission and one of the main reasons for rethinking the intersections between children, education, and geography.

Due to the different heritages, histories, and disciplinary backgrounds, geography education, geographies of education, and children's geographies might each be considered examples of what Lave and Wenger (1991) term 'communities of practice' (Finn et al., 2021). This conceptualisation is helpful in considering the different barriers individuals and communities might face when engaging with one another, and the impacts this has on research and teaching (Ibid.). For example, a person studying geography education as part of their initial teacher education programme may never be taught about geographies of education. This may lead to them being under-informed about the social and spatial injustices children, young people, and communities face and how these impact on teaching, learning, and day-to-day life in their placement school.

However, over recent years, exciting and significant new connections between geography and education have evolved (Puttick, 2022). For example, the Royal Geographical Society's Higher Education Research Group (HERG) reformed in 2019, as the Geography and Education Research Group (GeogEd). As Healey et al. (2020: p. 12) explain, the motivation for this change was to:

> Re-invigorate the connections between different levels of geography education to focus on a research-informed, discipline-based approach to staff development and the enrichment of teaching and the curriculum through the twin foci of geography education and geographies of education.

They suggest that these connections have the potential to support and inform research agendas and staff development. Here, greater engagement with the intersections between children, education and geography offers opportunities to inform and enrich research and practice in many areas including, but not limited to:

- supporting academic and social transitions for students as they move between educational phases (Tate and Hopkins, 2019; Biddulph et al., 2022);
- investigating and attending to social and spatial injustices in education and educational spaces (Taylor, 2009);
- ensuring that children's geographies are recognised and valued in education (Catling and Martin, 2011; Hammond, 2022);
- exploring the relationships between educational institutions, communities, and wider public services and/or infrastructure such as housing (Taylor, 2009);
- supporting and informing curriculum design at a multitude of scales (Morgan, 2022);
- informing how ideas move and are 'recontextualised' between disciplines and subjects (Finn, 2021);

- facilitating the development of research projects and developing the research practices and research 'literacy' of teachers (Healy, 2022; Mitchell and Béneker, 2022); and
- supporting and informing pedagogy and teaching (Roberts, 2017).

Indeed, as Norcup (2015: p. 31) explains when discussing children's geographies and geographies of education, these subdisciplines 'potentially offer increased space across and through which academic geographers are able to engage with the myriad geographies connected with educational processes and places'.

However, these outcomes are neither inevitable nor guaranteed. For example, whilst Morgan (2022) presents a picture of engagement with children's geographies – in terms of both children's own experiences and imaginations of the world and the subdiscipline – throughout the recent history of school geography in England, other research suggests that they have been marginalised and 'pushed out' by policies and practices related to accountability and performativity (Hammond and McKendrick, 2020) and conceptualisations of which/whose knowledge is 'powerful' (Catling and Martin, 2011). As such, in rethinking the intersections between children, education, and geography, it is important to consider not only the value of potential research across these areas, but also the structures and systems which support and inform interaction and how best to operate within and around these. We now move on to introduce the three sections and ethical practices which frame the book.

Introducing the book: framing, language, and ethics

This book is divided into three interrelated sections. Each section makes a distinct contribution to our objective of rethinking the intersections between children, education, and geography to better understand how injustices are (re)produced in, and through, education and to consider how we can best support and empower children and young people through (geography) education in their lives and futures.

Section I: geographies of education and educational spaces comprises reflections on the multi-scalar nature of education spaces and how these relate to the lives of those who work and study within them, and the places they exist within. Here, the authors consider how geography can provide unique insights into formal and alternative educational spaces. In Chapter 2, *geographies of education at macro-, meso-, and micro-scales: young people and international student mobility*, Johanna Waters and Rachel Brooks focus on international student mobility for educational opportunities and examine the ways students and places shape, and are shaped by, education. In Chapter 3, *geographies of education spaces: architecture, materialities, power and identity*, Peter Kraftl focuses on the micro scale of educational spaces to consider why these spaces matter beyond the often-dominant focus on student outcomes. The final chapter in this section, Chapter 4, by John H. McKendrick, *children's geographies and schools: beyond the mandated curriculum*, considers geography at a curriculum scale, arguing that school geography has the potential to enable children and young people to understand and confront social injustices that frame their lives.

Underpinning this section is active consideration of power; the power of young people to shape educational spaces at a range of scales; power dynamics and positionalities which are constructed and enacted in formal and informal spaces of education, and how these impact on children's and young people's identities and experiences; and the power of geography education beyond the mandated curriculum to expose and to challenge deep-seated social inequalities that are often reproduced through education.

Section II: children's geographies and their significance in, and to, everyday life and education comprises five chapters that examine the realities, the practicalities and the importance of children's geographies in, and to, education and society. In Chapter 5, *connecting children and young people's geographies and geography education: why this matters to and for children, education and society*, the authors, Mary Biddulph, Peter Hopkins and Simon Tate, utilise two case studies from research with young people to consider the potential of a mutually enriching relationship across research in young people's geographies, school geography and education more generally. Chapter 6, by Simon Catling and Susan Pike, *becoming acquainted: aspects of diversity in children's geographies* calls for younger children to have greater agency over their learning, as their geographies afford them unique insights into matters of concern from a local to a global scale. The theme of voice and agency is picked up in Chapter 7 by Lauren Hammond and Grace Healy in their chapter *student voice, democratic education, and geography: reflecting on the findings of a survey of undergraduate geography students*. Here, the views of undergraduate students on their geographical education to date support calls for more democratic approaches to geographical teaching and learning to ensure educators have a better understanding of the children and young people who they teach. Chapter 8, *the value of geography to an individual's education* by David Lambert and Kelly León, considers the value of geographical thinking to children and young people as individuals. In the chapter, Lambert and León present an approach in California, whereby young people were able to exercise greater personal influence over the geography they learnt. The final chapter in this section, Chapter 9 by John Morgan, *young people's geographies, schooling, and the curriculum problem: where have all the cool places gone?* uses the text *cool places: geographies of youth cultures* edited by Tracey Skelton and Gill Valentine (1998) as a starting point from which to critically examine the changing nature of young people's agency and the contribution of geographical thinking to this. The chapter challenges the reader to consider how school geography can again connect with young people's experiences at a time when the official curriculum seeks to do otherwise.

Overall, this section explores the intrinsic relationships between children's everyday geographies and the potential of these geographies to shape, inform, and improve curriculum and pedagogy and ultimately the educational experiences of children and young people. In diverse ways, the chapters explore the complexity of children and young people's lives as a means of encouraging educators to truly engage with the children they teach.

Section III progressive geographies in education is the last section and comprises five chapters. These critically consider what a progressive geographical education might look like. The aim here is to consider how geography education can better support young people in making sense of the world they live in and contribute to, both today and in the future. The chapters capture ways in which geography as an academic discipline, school geography, and children's lives intersect at a curriculum and pedagogical level. The first chapter in this section, Chapter 10 *de/colonising the (geography) curriculum* by Fatima Pirbhai-Illich and Fran Martin, sets the scene by posing some challenging but essential questions about what gets taught in school geography, who decides this, and what gets excluded and why. Subsequent chapters then build on these questions and consider, in different ways, what a more progressive school geography might look like. Chapter 11, *climate change education: following the information* by Steve Puttick, Paloma Chandrachud, Rahul Chopra, James Robson, Sanjana Singh, and Isobel Talks uses 'story telling' as a means of examining the challenges of teaching climate change in schools and the challenge for teachers and students to critically engage with the 'superabundance' of information and misinformation. The authors consider the ways certain knowledge and information are privileged over others and the consequences of this for young people's real engagement with the consequent injustices of climate change. Chapter 12, *expanding students concept of 'home': teaching migration with a geographical capabilities approach,* by David Mitchell and Tine Benéker focuses on the ways a GeoCapabilities approach to curriculum thinking enabled teachers in England and the Netherlands to reconsider their teaching of migration. By focusing on the concept of 'home' and the notion of 'homemaking' teachers felt they were better able to utilise the everyday and personal geographies of young people, including their students, to challenge the more stereotypical ideas about migration often taught in schools. Chapter 13 by Ria Dunkley *looking closely for environmental learning: citizen science and environmental sustainability education* reconsiders the intersections between formal and informal learning spaces through children and young people's participation in citizen science. The chapter presents the work of the 'Spot a Bee' project to illustrate the potential of transdisciplinary learning in creating an 'ecopedagogy of hope' as a counterbalance to the often-overwhelming message of climate catastrophe. Chapter 14, the final chapter in this section, is by Helen Clarke and Sharon Witt and is titled *paying attention with more-than-human worlds: field-visiting.* Through the use of poetic vignettes, Clarke and Witt examine the notion of 'field-visiting' and its associated 'pedagogies of attention' as the means by which people of all ages can better understand, appreciate, and live with the more-than-human world. Ultimately, this section is framed by the notion of justice and consideration of what this means in, and for, education and specifically its relationships to geography education.

The book concludes by reflecting on the extent to which we have addressed our objectives and looks towards a future enriched by work at the intersections of children's geographies, geography of education, and geography education. Before we conclude this chapter, we offer a note on the use of language in the book and reflect on ethics as a fundamental underpinning of research across the intersections between children, education, and geography.

As the term *children's geographies* refers to both children's geographies in and of the world, and the subdiscipline of geography, the authors clarify to which they are referring in their chapter. Unless explicitly stated by chapter authors, we also differentiate between children's geographies and young people's geographies. *Children's geographies* refer to those under 11/12 years of age, and *young people's geographies* refers to those between 11/12 and 25 years of age. Whilst we recognise this is an artificial division, as the book focuses on education, the division is helpful to clarify differences across primary, secondary, and tertiary education. Finally, we use the terms *teacher of geography* (as opposed to geography teacher), which better reflects the professionals who teach geography (not all of whom specialise in teaching geography, in particular in primary school settings). The term *geography teacher educator* refers to a person working in teacher education in contexts including schools and universities.

Many of the chapters here report the outcomes of educational research, sometimes research involving young people. As such, it is important to be explicit about the ethical dimension of our work. As Wellington (2000; p. 54) explains:

> An 'ethic' is a moral principle or a code of conduct which . . . governs what people do. It is concerned with the way people act or behave. The term 'ethics' usually refers to the moral principles, guiding conduct, which are held by a group or even a profession (though there is no logical reason why individuals should not have their own ethical code).

There are ethical behaviours a society has a right to expect from, for example, its teachers or its doctors or its legal system (Committee on Standards in Public Life, 1995). In the context of educational research, Wellington reminds us that ethics is not merely a set of processes to complete in order to gain institutional approval for research to take place. Rather, it refers to the attitudes and behaviours that those 'outside' research have a right to expect of those conducting research and implies that these attitudes and behaviours should permeate all aspects of research activity. As a group of editors, it has been important to us that, from the outset, the development, construction, and writing of this book are consistent with the highest standards of ethical practice.

In conclusion . . .

On Blake's memorial in the crypt of St Paul's Cathedral (London, England) is written the first line of another of his poems 'Auguries of Innocence':

> To see a World in a Grain of Sand
> And a Heaven in a Wild Flower
> Hold Infinity in the palm of your hand
> And Eternity in an hour.

Blake reminds us of the wonders of life and the Earth, and of the pleasures and endless possibilities of innocence and imagination. In the poem, Blake goes on to write about the delicate relationships between people, nature, and the Earth, of inequities in society, and encourages us to reflect on how we act in the world as individuals, and the societies and places we create. Here Blake, rather like Haraway's work many years later, could also be seen to be asking us to *stay with the trouble*, to be truly present, to connect with the moment, and to think about our positionalities as people in the world. As you read this edited collection, we encourage you to reflect on the possibilities and potential of rethinking the intersections between children, education, and geography; what this might mean for you and your context but more broadly for children, education, society, and the Earth.

References

Aitken, S. (2018a) *Young People, Rights and Place: Erasure, Neoliberal Politics and Postchild Ethics.* Abingdon: Routledge

Aitken, S. (2018b) 'Children's geographies: Tracing the evolution and involution of a concept' *Geographical Review* 108(1) pp3–23

Biddulph, M. Hopkins, P. Tate, S. (2022) 'Connecting children's and young people's geographies and geography education: Why this matters to and for children, education and society' in Hammond, L. Biddulph, M. Catling, S. McKendrick, J. H. (eds.) *Children, Education and Geography: Rethinking Intersections.* Abingdon: Routledge

Blake, W. (1789) *'The School Boy' (A poem by William Blake) Reproduced in Songs of Innocence and Experience: With an Introduction and Commentary by Sir Geoffrey Keynes.* Oxford: Oxford University Press

Brock, C. (2016) *Geography of Education: Scale, Space and Location in the Study of Education.* London: Bloomsbury Academic

Brooks, R. Waters, J. (2017) *Materialities and Mobilities in Education.* Abingdon: Routledge

Butt, G. (2019) *Geography Education Research in the UK: Retrospect and Prospect: The UK Case within the Global Context.* Cham: Springer

Catling, S. (1988) 'Children and geography' in Mills, D (ed.) *Geographical Work in Primary and Middle Schools.* Sheffield: Geographical Association, pp. 9–18.

Catling, S. Martin, F. (2011) 'Contesting powerful knowledge: The primary geography curriculum as an articulation between academic and children's (ethno-) geographies' *The Curriculum Journal* 22(3) pp317–335.

Catling, S. Pike, S. (2022) 'Becoming acquainted: Aspects of diversity in children's geographies' in Hammond, L. Biddulph, M. Catling, S. McKendrick, J. H. (eds.) *Children, Education and Geography: Rethinking Intersections.* Abingdon: Routledge

Catling, S. Willy, T. (2018) *Understanding and Teaching Primary Geography.* London: Sage.

Committee on Standards in Public Life (1995) *The Seven Principles of public life.* Available at: *www.gov.uk/government/publications/the-7-principles-of-public-life/the-7-principles-of-public-life – 2* (Accessed 7th February 2022)

Department for Education [DfE]. (2021) *Policy Paper: Initial Teacher Training Market Review: Overview.* London: DfE. Available at: www.gov.uk/government/publications/initial-teacher-training-itt-market-review/initial-teacher-training-itt-market-review-overview (Accessed 8th February 2021).

Department for Education [DfE]. (2022) *Political Impartiality in Schools.* London: DfE. Available at: www.gov.uk/government/publications/political-impartiality-in-schools/political-impartiality-in-schools (Accessed 12th February 2022)

Dolan, A. (ed.) (2022) *Teaching Climate Change in Primary Schools*. Abingdon: Routledge.

Dunkley, R. (2022) 'Looking closely for environmental learning: Citizen science and environmental sustainability education' in Hammond, L. Biddulph, M. Catling, S. McKendrick, J. H. (eds.) *Children, Education and Geography: Rethinking Intersections*. Abingdon: Routledge

Finn, M. (2021) 'Questioning recontextualisation: Considering recontextualisation's geographies' in Fargher, M. Mitchell, D. Till, E. (eds.) *Recontextualising Geography*. Switzerland: Springer

Finn, M. Hammond, L. Healy, G. Todd, J. Marvell, A. McKendrick, J. H. Yorke, L. (2021) 'Looking ahead to the future of GeogEd: Creating spaces of exchange between communities of practice' *Area*. https://doi.org/10.1111/area.12701

Freire, P. (1984) *The Politics of Education: Culture, Power and Liberation*. Westport: Bergin and Garvey

Freytag, T, Lauen, D., Robertson, S (eds.) (2022) *Space, Place and Educational Settings*. Cham: Springer

Giroux, H. (1982) 'The politics of educational theory' *Social Text* 5 pp87–107

Hammond, L. (2022) 'Recognising and exploring children's geographies in school geography' *Children's Geographies* 20(1) pp64–78. DOI: https://doi.org/10.1080/14733285.2021.1913482

Hammond, L. McKendrick, J. H. (2020) 'Geography teacher educators' perspectives on the place of children's geographies in the classroom' *Geography* 105(2) pp86–93

Haraway, D. (2016a) *Staying with the Trouble: Making Kin in the Chthulucene*. Durham: Duke University Press

Haraway, D. (2016b) 'Staying with the trouble: Anthropocene, capitalocene, chthulucene' in Moore, J. W. (eds.) *Anthropocene or Capitalocene? Nature, History and the Crises of Capitalism*. Oakland: PM Press

Healey, R. France, D. Hill, J. West, H. (2020) 'The history of the Higher Education Research Group of the UK Royal Geographical Society: The changing status and focus of geography education in the academy' *Area*. DOI: 10.1111/area.12685

Healy, G. (2022) 'Geography and geography education scholarship as a mechanism for developing and sustaining mentors' and beginning teachers' subject knowledge and curriculum thinking' in Healy, G. Hammond, L. Puttick, S. Walshe, N. (eds.) *Mentoring Geography Teachers in the Secondary School: A Practical Guide*. Abingdon: Routledge

Holloway, S. Hubbard, P. Jöns, H. (2010) 'Geographies of education and their significance to children, youth and families' *Progress in Human Geography* 34(5) pp583–600

Hopkins, P. (2010) *Young People, Place and Identity*. Abingdon: Routledge

Horton, J., Kraftl, P., Tucker, G. (2008) 'The challenges of 'children's geographies': A reaffirmation' *Children's Geographies* 6(4) pp335–48.

Janhnke, H., Kramer, C., Meusburger (eds.) (2019) *Geographies of Schooling*. Cham: Springer.

Kavanagh, A., Waldron, F., Mallon, B. (eds.) (2021) *Teaching for Social Justice and Sustainable Development across the Primary Curriculum*. Abingdon: Routledge

Kraftl, P. (2021) *Slow Violence: A Reimagining Childhood Webinar*. Available at: www.youtube.com/watch?v=qyAl21PMRYA (Accessed 6th January 2022)

Kraftl, P. Andrews, W. Beech, S. Cesera, G. Holloway, S. L. Johnson, V. White, C. (2021) 'Geographies of education: A journey' *Area*. https://doi.org/10.1111/area.12698

Lambert, D. (2010) 'Geography education research and why it matters' *International Research in Geographical and Environmental Education* 19(2) pp83–86

Lambert, D. León, K. (2022) 'The value of geography to an individual's education' in Hammond, L. Biddulph, M. Catling, S. McKendrick, J. H. (eds.) *Children, Education and Geography: Rethinking Intersections.* Abingdon: Routledge

Latour, B. (2017) *Facing Gaia: Eight Lectures on the New Climatic Regime.* Cambridge: Polity Press

Latour, B. (2018) *Down to Earth: Politics in the New Climatic Regime* (Translated by Catherine Porter). Cambridge: Polity Press

Lave, J. Wenger, E. (1991) *Situated Learning: Legitimate Peripheral Participation.* Cambridge: Cambridge University Press

Mills, S. (2022) *Mapping the Moral Geographies of Citizenship: Character, Citizenship and Values.* Abingdon: Routledge

Mitchell, D. Béneker, T. (2022) 'Expanding students' concept of 'home': Teaching migration with a geographic capabilities approach' in Hammond, L. Biddulph, M. Catling, S. McKendrick, J. H. (eds.) *Children, Education and Geography: Rethinking Intersections.* Abingdon: Routledge

Morgan, J. (2022) 'Where have all the cool places gone? Young people's geographies, schooling and the curriculum problem' in Hammond, L. Biddulph, M. Catling, S. McKendrick, J. H. (eds.) *Children, Education and Geography: Rethinking Intersections.* Abingdon: Routledge

Nixon, R. (2011) *Slow Violence and the Environmentalism of the Poor.* Cambridge: Harvard University Press

Oswell, D. (2013) *The Agency of Children: From Family to Global Human Rights.* Cambridge: Cambridge University Press

Pini, B. Gulson, K. N. Kraftl, P. Dufty-Jones, R. (2017) Critical geographies of education: An introduction. *Geographical Research* 55 pp13–17

Puttick, S. (2022) 'Geographical education I: Fields, interactions and relationships' *Progress in Human Geography.* DOI:10.1177/03091325221080251

Roberts, M. (2017) 'Geography education is powerful if . . .' *Teaching Geography* 42(1) pp6–9

Rousell, D. Cutter-Mackenzie-Knowles, A. (2020) 'A systematic review of climate change education: giving children and young people a 'voice' and a 'hand' in redressing climate change' *Children's Geographies* 18(2) pp191–208

Rushton, E. Dunlop, L. Atkinson, L. Price, L. Stubbs, J.E. Turkenburg-van Diepen, M. Wood, L. (2021) 'The challenges and affordances of online participatory workshops in the context of young people's everyday climate crisis activism: insights from facilitators' *Children's Geographies.* DOI: 10.1080/14733285.2021.2007218

Skelton, T. Valentine, G. (eds.) (1998) *Cool Places: Geographies of Youth Cultures.* London: Routledge.

Skovdal, M. Benwell, M. C. (2021) 'Young people's everyday climate activism: New terrains for research, analysis and action' *Children's Geographies* 19(3) pp259–266

Standish, A. (2021) 'Geography' in Seghal Cuthbert, A. Standish, A. (eds.) *What Should Schools Teach: Disciplines, Subjects and the Pursuit of Truth.* London: UCL Press

Tate, S. Hopkins, P (2019) 'Student perspectives on the importance of both academic and social transitions to and through their undergraduate degree' in Walkington, H. Hill, J. Dyer, S. (eds.) *Handbook for Teaching and Learning in Geography.* Cheltenham: EE Publishing Limited.

Taylor, C. (2009) 'Towards a geography of education', *Oxford Review of Education* 33(5): 651–669

Walker, C. (2021) ''Generation Z' and 'second generation': An agenda for learning from cross-cultural negotiations of the climate crisis in the lives of second generation immigrants' *Children's Geographies* 19(3) pp267–274

Walshe, N. (2017) 'An interdisciplinary approach to environmental and sustainability education: Developing geography students' understandings of sustainable development using poetry' *Environmental Education Research* 23(8) pp1130–1149. DOI: 10.1080/13504622.2016.1221887

Walshe, N. Sund, L. (2022) 'Developing (transformative) environmental and sustainability education in classroom practice' *Sustainability*, 14(1) p110 https://doi.org/10.3390/su14010110

Waters, J. (2018) *Geographies of Education: Oxford Bibliographies*. Available at: www.oxford-bibliographies.com/view/document/obo-9780199874002/obo-9780199874002-0182.xml (Accessed 28th February 2022)

Wellington, J. (2000) *Educational Research: Contemporary Issues and Practical Approaches*. London: Continuum

West, H., Hill, J., Finn, M., Healey, R.L., Marvell, A. Tebbett, N. (2020) 'GeogEd: A new research group founded on the reciprocal relationship between geography education and the geographies of education' *AREA*. DOI: 10.1111/AREA.12661.

Yusoff, K. (2018) *A Billion Black Anthropocenes or None*. Minneapolis: University of Minnesota Press

SECTION I

Geographies of education and educational spaces

2

GEOGRAPHIES OF EDUCATION AT MACRO-, MESO-, AND MICRO-SCALES

Young people and international student mobility

Johanna Waters and Rachel Brooks

Introduction

> Although the nation state is still the implicit spatial framework for research on student migration, many of the explicit comparisons that students engage in appear to be in relation to places.
>
> (Raghuram, 2013: p. X)

The 'where' of education – and the role played by mobilities in this process – is increasingly of interest to researchers (Finn and Holton, 2019; Waters, 2017). Education is now understood not as something that happens 'in [a] place' but instead unfolds over multiple spaces and times. Furthermore, education is enlisted in the *creation* of space (Massey, 2005), signalling the coming together of coterminous relational processes in diverse constellations, be that the classroom, lecture theatre, school bus, residential block, city, or even airport. Mobilities and materialities in education are co-constitutive (Brooks and Waters, 2017).

We use the international student mobility (ISM) of higher education students (at the upper end of the age range for work on children and young people's geographies) as a lens through which to explore the different scales at which education could be said to be 'occurring'. The concept of ISM predictably foregrounds the 'international' scale (Brooks and Waters, 2022). In fact, however, its processes and impacts can be observed at *multiple scales* at the same time. Consequently, ISM provides an exemplar of the complexity of the geographies of education: how they represent the 'product of interrelations', 'constituted through interactions, from the immensity of the global to the intimately tiny' (Massey, 2005: p. 9). ISM also exemplifies the claim that space represents 'contemporaneous plurality' – a 'sphere' of 'coexisting heterogeneity' (Ibid.). Although we focus on older 'young people' here (higher education students), similar issues might be highlighted for work on

DOI: 10.4324/9781003248538-4

younger (pre-school and school-age) children seeking education across borders (Leung and Waters, 2021), and we draw out these links when we can. We proceed, now, with a discussion of ISM at different scales – from the macro to the micro. After Massey (2005), in conclusion we consider the implications of such a perspective for the geographies of education (see Waters and Brooks, 2021).

ISM and 'international' space

Whilst learning is understood to 'take place' within particular, identifiable and 'local' places (Waters, 2017), what we could call the spaces of education are far broader and more diverse than this depiction would suggest. There is, as we discuss later, an international/global 'space' of education. Despite being largely symbolic in nature, the perpetual construction of this international space of education matters as it is directly implicated in where international students choose to study and why. Macro-level depictions of education are powerful and are underpinned by long-standing colonial and imperial relationships. It is to these that we first turn our discussion.

The OECD (2018) has described the increase in the total number of international students (i.e., tertiary level students studying for undergraduate, master's, and doctoral degrees) over the past several decades as 'rising from 2 million in 1999 to 5 million 17 years later' (p. 219).[1] These numbers signal a significant 'global' population of mobile young people. As the OECD suggest in their report, it is possible to visualise this mobility in 'macro scale' terms:

> Students from Asia form the largest group of international students enrolled in OECD tertiary education programmes at all levels (1.9 million, 55 per cent of all international students in 2016. . . . Of these, over 860 000 come from China. Two-thirds of Asian students converge towards only three countries: Australia (15%), the United Kingdom (11%) and the United States (38%).
>
> *(OECD, 2018: p. 221)*

Such grey literature tends to divide the world up into regions: Asia is the most significant 'sender' of international students and China the biggest 'source' country. Despite a diversifying of student destinations (as will be described later), countries located within the Anglophone 'West' continue to attract the most internationally mobile young people. The OECD (2019) highlights that 'English-speaking destinations' remain the most attractive to international students, and students from Asia represent the majority of international students globally. The report accounts for these geographical patterns by arguing that 'domestic' and 'external' factors combine to create the 'push' and 'pull' of international study. Knowledge-based economies around the world require skilled workers leading to an overall increase in demand for tertiary education. Local education systems cannot always meet that demand. In addition, the report notes the importance of cheaper flights and the spread of the internet, making ISM not only 'more affordable' but also 'less irreversible' than previously, as students engage in forming transnational ties that represent both macro- and micro-scales (Waters and Brooks, 2021).

Interpreting these broad, macro-trends often relies on recourse to an enduring legacy of colonialism and imperialism. The English language is a key factor shaping these geographies. International students from parts of Asia may draw on 'historically rooted imaginative geographies in which the "West" is seen to be more sophisticated and advanced than the "East"' (Kölbel, 2020: p. 96), often resorting to stereotyped imaginaries (Rose-Redwood and Rose-Redwood, 2019). International league tables have reinforced these ideas: until relatively recently, universities within North America and the UK have dominated such international rankings – league tables are often an important source of 'objective' information on the value of particular universities for international students (Jöns and Hoyler, 2013), and some international scholarships depend on an institution's ranking (see arguments in Waters and Brooks, 2021).

Macro-level patterns of ISM can also be related to the 'supply side' (which is often, in the literature, overlooked in comparison to student 'demand'). This can include, for example, state policy with respect to international students, as discussed in Findlay's (2011) work. This might involve immigration policy (Geddie, 2015; Robertson, 2011), access to post-study work visas, and the potential for permanent residency following graduation, all of which can appear within statewide marketing campaigns to attract international students (Sin et al., 2021). There is also a significant role to be played by various intermediaries in directing students to 'dominant' countries. These might include educational organisations such as (in the context of the UK) Universities UK and the British Council (Findlay, 2011; see also Sidhu, 2006; Beech, 2019), which promote certain destinations (and educational institutions). Further research has also emphasised the important influence of educational agents and consultants in channelling students to particular destinations. Such consultants tend to draw upon the crude, macro-geographical representations mentioned earlier (Altbach, 2013; Beech, 2019; Collins, 2012; Kölbel, 2020; Nikula and Kivistö, 2020). Thus, whilst macro representations of ISM can be, to a certain extent, misleading and homogenising, presenting narrowly drawn assumptions about international student origins and destinations, they are also, as shown, rather powerful depictions that hold sway within the global higher education field.

Emergent macro-spaces of international students

Over the past decade, international league tables have charted some changes in these macro-scale geographies. In 2019, seven Asian universities appeared in the top 50 of 'global universities' according to the *Times Higher Education* World University Rankings. University World News (2020) recently wrote:

> Asian universities have gained ground with record representation in the top 100, while the United States, United Kingdom and other European countries

> have seen an overall decline in performance . . . This year's instalment of the rankings sees Asian universities enjoy a greater presence among the global top 100 than at any point in the ranking's 17-edition history. There are now 26 Asian universities achieving top-100 ranks.
>
> *(n.p.)*

China is now the world's third largest recipient country of international students, after the United States and the UK (Moe, 2019).

This represents a significant shift, over the past few decades, in how universities outside the West have been perceived, evaluated, and represented and, relatedly, how regions such as East Asia have sought actively to promote themselves as attractive *destinations* for international students. Sidhu et al. (2019) discuss what they term 'Asian regionalism' in relation to ISM, adopting a post-colonial lens. This discussion includes a consideration of 'the governmentalities of a region described as "Rising Asia"'; in other words, how states within Asia are actively attempting to counter narratives of Western superiority with various state-level policies and strategies. Many of these relate directly to higher education[2] and attempt favourably to position Asian universities within global knowledge and innovation networks. Asian regionalism in part reflects deliberate attempts by non-Western states 'to assemble and govern "new knowledge spaces" in and through the institution of the "world-class" university' (Sidhu et al., 2019: p. 31). In other words, universities have become central to states' economic strategies based on the ascendance of the idea of the knowledge economy. Such an approach, in its attempts to counter Western-centric depictions of international education, is part of a wider movement towards decolonisation within academia (Jazeel, 2019; Müller, 2021). Although research on ISM has been surprisingly slow to engage within post-colonial approaches, there are several key examples to which we can now point, including Madge et al. (2009), Raghuram et al. (2020), França et al. (2018), Riaño et al. (2018), and Yang (2020). In recent scholarship, there has been more explicit recognition that international mobility itself can reinforce and entrench pre-existing (post) colonial structures (Koh, 2017; Sin, 2009), representing a macro-scale of analysis.

The meso-scale: cities and international students

To date, the role that cities play in ISM has been largely overlooked. Cities have been perceived primarily as 'destinations' of international students (e.g., Collins, 2008, 2010; Fincher and Shaw, 2009; Ma, 2020) despite the ways in which young people undoubtedly transform urban landscapes through their consumption practices and residential choices (Smith and Hubbard, 2014; Collins and Ho, 2014). Furthermore, only a handful of projects have considered the part that cities can play in creating particular flows of ISM. Our arguments have been influenced by Collins's (2014) claims about the 'contingent assembly of the urban and its role in globalising higher education' (p. 242), wherein cities are perceived as more than a mere 'backdrop' to ISM.

Related to the imaginative geographies discussed earlier for the macro-scale, similar geographical imaginations come into play when we consider the role of cities. Cities are often framed as 'desirable places', inducing 'aspirations to become mobile' amongst international students (Collins, 2014: p. 243). They can also, as Collins has argued, be important in students' learning of 'place-specific knowledge' (p. 243). Moreover, researchers interested in how 'prestige' is created within ISM have begun to explore how this might intersect with the identities of cities (Findlay et al., 2012; Beech, 2014). Prazeres et al. (2017), for example, argue that the 'distinctive qualities' of a place might attract international students to 'less prestigious' higher educational institutions, assuming that the city aligns with the 'specific lifestyle desires' of students and their 'personal imaginaries' (p. 116).

For students with no direct experience of study abroad, the imaginative geographies that attach to cities can play a key role in their mobilities (Beech, 2014, Kölbel, 2020). Kölbel (2020) explored the media's influence in shaping the ideas of 40 young people from Kathmandu, Nepal around ISM. Young people's choice of 'destination city', Kölbel argues, was fundamentally influenced by the media, which has created 'dominant imaginaries attached to international student mobility [which] shape young people's identities and their life chances' (2020: p. 88). Young people in Kathmandu are consistently exposed (by educational marketing campaigns) to images and ideas relating to study abroad, including 'countless billboards' advertising study abroad opportunities and advertisements in daily newspapers and flyers. Significantly, Kölbel (2020) notes, exposure to such images by individuals rarely results in a fuller understanding of overseas places but rather creates a partial and incomplete picture of particular cities and what it would be like to study there.

In a somewhat different project, we discussed the role of the city (in this case, London) in the development of so-called satellite or branch campuses by UK universities. These began to appear around 2009 and, by 2014, there were 13 in operation (QAA, 2014). Although a small number of these campuses have since shut down, it was recently announced that the University of Portsmouth will soon commence opening a campus in London (BBC, 2021): there remains an appetite for such developments within UK higher education (see Brooks and Waters, 2018 for a fuller discussion of this). The position of London, as a pre-eminent global city, is crucial to this story (Sassen, 1991). UK higher education institutions appeal to and draw upon such place-making (Beech, 2019) to attract international students to branch campuses. The decision to locate branch campuses in London is very deliberate, directly linked to their appeal to (international) students. Such appeal includes (but is not limited to) aesthetic elements (high-rise buildings and iconic landmarks); its global financial district (concentrating headquarters of banks and other financial and legal institutions); transportation and communications connectivity; cultural appeal (Britishness and a cosmopolitan internationalism); leisure and arts facilities (theatres, cafes, restaurants and night life); and 'prestigious' higher education institutions centred on the colleges of the University of London (including King's, Imperial, London School of Economics and UCL).

International students have also been linked to the process of 'studentification' in the UK (Smith, 2005; Smith and Holt, 2007) and elsewhere, including in Ireland (Kenna, 2011), China (He, 2015), and Spain (Garmendia et al., 2012). Revington and August (2020) have recently discussed one aspect of this: the financialisation of the student housing market in Canada (the growth of the so-called Purpose-Built Student Accommodation (PBSA) since 2011). PBSA has also been discussed in the context of housing markets in the UK and Australia (Nakazawa, 2017). Here, the differences between international and domestic students are often highlighted: in Australia and New Zealand, for example, domestic students will habitually remain in the family home and commute to university whilst international students are channelled into PBSAs via university housing offices or private housing companies catering to students (Nakazawa, 2017). Consequently, the segregation of international students in city spaces becomes apparent. Fincher and Shaw (2009, 2011) observed this is not just for housing but also for leisure and cultural spaces. Often, universities (unintentionally, through their practices) end up spatially separating international students and domestic students, with implications for how diversity within the institution and the broader urban area is experienced (Fincher and Shaw, 2009) – linking the meso- and micro-scales, discussed later.

Calvo (2018) argues that international students have an even wider role to play in changing the fabric of cities and suggests that 'as a new class of transnational urban consumers [their impact] has been widely disregarded' (p. 2143). Their role as consumers in part relates to a perception than many international students are wealthy and have money to spend on leisure, including eating out and consumer goods. Collins (2008), for example, in studying ISM in Auckland, New Zealand, has discussed the proliferation of 'Korean' restaurants as a direct consequence of ISM.

Another meso-scale outcome of ISM is the development of urban education hubs: purpose-built areas of cities that focus specifically on the provision of (international) education (Kleibert, 2022). As Knight (2018) describes: 'the term education hub is used by countries seeking to position themselves as centers for student recruitment, education and training, research, and innovation' (p. 638). They are, in other words, state-level developments undertaken at a city-scale – drawing on the concentration of resources, facilities, and opportunities characteristic of large urban areas. The most well-known examples of urban education hubs are found in the United Arab Emirates, Qatar, Botswana, Malaysia, Singapore, and Hong Kong (Knight, 2018), whilst lesser-known examples continue to develop around the world: 'for instance Panama City, Bangalore in India, and Monterey in Mexico . . . have been seeking to brand themselves as education or knowledge cities' (Knight, 2018: p. 643). These hubs are used by domestic students but primarily attract international students from the wider region and beyond. It is instructive to note, as Knight (2018) has done, that education hubs have not been developed in countries deemed 'popular' destinations for international students (for example, the US or the UK). Instead, smaller countries outside of the traditional ISM nexus have used these as a draw. For some smaller countries and city-states, attracting international students has become an important pillar of economic development. Singapore was renowned for its urban hub initiative in the early 2000s, which developed an alliance with MIT (Massachusetts Institute of Technology) in a bid to become the 'Boston of the East' (Sidhu et al., 2011; Collins et al., 2014). The policy was named the Global School House and, in a study of this policy, Sidhu et al. (2011) set out to understand what attracted international students to Singapore (and if these considerations aligned with the policy). They concluded overwhelmingly that students valued Singapore's safe urban environment, followed by proximity to friends and family (presumably residing in the wider South East Asian region). Also, they observed: 'more than a third of those surveyed mentioned good job prospects as a reason for coming to Singapore [which] suggests that the government's promotional messages of an economically dynamic global city in a bustling region is being heard' (Sidhu et al., 2011: p. 35–36).[3] To date, however, little research has explored how students *experience* urban education hubs – what it is like to study and live within one and what their subsequent outcomes might be.

This subsection has drawn out the quite diverse 'urban geographies' of international higher education and student mobilities, discussing consumption, residential segregation, and urban hubs at a meso-scale. This scale is generally neglected in the wider literature and discussions around ISM – yet is crucial to the ways in which students engage with ISM, in terms of what attracts them to particular places and how they experience them. We now turn to consider the micro-scale, after Hall's (2019) extensive efforts to highlight the political and ethical significance of foregrounding the everyday in discussions of social geographies. We do this through a focus on the university campus and the role it plays in the social lives of international students.

Micro-scale geographies of ISM

Through their scholarship, Holton and Riley (2013) have highlighted the everyday, on-campus experiences of university students (Anderson et al., 2012; Hopkins, 2011). The campus and its associated classrooms, seminar rooms, lecture theatres, and cafes can represent an important space for international students, as Chacko (2020) writes in relation to international students in Singapore: 'Campus was the place where students felt most at home' (p. 12). International students' quotidian experiences and social interactions often unfold within the university campus space. Lee (2020), for example, discussed the 'cafeteria food' eaten by international students in China, seen by the students to represent the cosmopolitan nature of their experiences. The space of the university was also discussed by Beech (2018), who argued that:

> Universities may do little to encourage cross-cultural and multi-cultural communication both inside and outside of the classroom . . . International students mentioned, for example, that they were housed separately from local or host students and one of the universities, at the time of the research, offered a separate welcome week programme for international students and

so from the outset they were treated as a distinctive other rather than as part of a unified student community.

(p. 23)

However, Beech also argues that *self*-segregation can be significantly beneficial to international students (and, indeed, chosen by them), offering them with the emotional and practical support that institutions often fail to provide. Students come to rely on the support networks they develop with co-nationals (Mittelmeier et al., 2018). Rose-Redwood and Rose-Redwood (2019: p. 25) discuss the perceived 'benefits of primarily interacting with co-nationals, including speaking a common language, eating similar food, abstaining from alcohol consumption and practising the same religion'. Usually, these activities will occur in particular spaces on, or within, university campuses, either face to face or in close proximity, representing the micro-scales of (international) education and sociality.

Friendships developed amongst international students can also function as social capital (Rose-Redwood and Rose-Redwood, 2019; Beech, 2018; Waters, 2009; Brooks and Waters, 2011) and these can be directed and influenced by the space of the classroom. Priya, an international student from India, cited in Beech's (2019) study, explained how the segregation of international students in the UK prevented the development of (some) friendships. She lived with other Indian students: international students at her university were housed separately. In class, she described how there is a 'European table, British table, there are two Indian tables and there's a Thai table and a Chinese table' (p. 21) and very little interaction between them.

Interestingly, however, attention to micro-scale geographies is often largely absent from work on ISM and social interactions. Little heed is paid to *where and how* interactions take place – the significance of space, spatiality, distance, and place-making (Brooks and Waters, 2017). Yuan (2011), for example, explored the experiences of Chinese students at an American university, highlighting the importance of 'socialisation': that is, 'hanging out' and living with other co-national students. However, the micro-geographies of these arrangements (where, for example, social activities took place and how living quarters were organised) were omitted: brief reference was made to how, when observing interactions on campus, 'It was not uncommon to see a Chinese student and an American standing next to each other and smiling at each other but having no conversations' (Yuan, 2011: p. 149–150), and yet the analysis fails to take the geographies of this encounter seriously. In contrast, we would like to suggest that social interactions (in the classroom, cafeteria, halls, or residence or lecture theatres) can be facilitated or significantly hindered by spatial factors, as Priya clearly indicated earlier (see Holton 2016a, 2016b on university halls and students' living arrangements). Such separation may, in fact, undermine any claims to a 'diverse' student body frequently made by universities in their marketing or advertising materials. Diversity should also, in theory, indicate spatial proximity and social interaction.

Sidhu et al. (2016) provide an interesting and unusual example of state intervention with regard to the social and spatial segregation of international students at a micro-scale. They compare different campus housing projects at three higher education institutions in East Asia. In some, international students were housed separately from domestic students, and whilst students often expressed 'approval' at these arrangements, this also prevented interactions 'that might generate uncomfortable but also generative socialities' between individuals (p. 11). At the National University of Singapore, for example, there was deliberate intervention to prevent damaging segregation and what they described as 'ghettoisation', by ensuring a 'good mix of students' within each hall of residence. This fitted, Sidhu et al. (2016) argued, with the institution's 'globalising vision', echoing the state's 'racialised governmentality':

> In short, rather than mobile students who are free to circulate as they desire, and perhaps encounter friction and risk in daily life, these universities have actively assembled campus spaces where flows of students are conducted to smooth their mobility and achieve strategic outcomes. While differing across the three campuses, these strategies unfold as practices of containment that are enacted through built form, imaginings of what it means to be global, research renderings of students and inducements towards diversity. International students, and particular nationalities, are defined as distinct and in need of management. The campus becomes demarcated in order to guide their bodies through educational and social spaces and to manage their integration into the broader urban fabrics the universities sit within.
>
> *(Sidhu et al., 2016: p. 11–12)*

Here, the link between the university campus and the wider city is highlighted, as is the role of the state (macro-scale) in managing international students' interactions on a micro- (and seemingly mundane) level.

Conclusion

Since the onset of the COVID-19 pandemic in 2020, the spaces of international students' (im)mobility have become more apparent than ever. International borders have been closed to international students, study abroad placements have been cancelled, and some students have found themselves either repatriated or 'stranded' overseas, unable to leave their accommodation. International students have experienced hotel quarantining and at least part of their learning has been transferred online. Above all else, the recent pandemic has brought into focus the *multiple* spaces of ISM and why they all matter. Students' everyday experiences of space have been transformed.

In this chapter, we stress the importance of thinking across spaces when trying to understand the nature of (international) education and how this relates to young people's experiences. Some media continue to depict a gigantic 'flow' of young people from the Global South to the Global North or from 'East' to 'West': these depictions and representations of international education continue to exert power.

Large-scale data sets collated by international organisations (such as the OECD) represent and reinforce dominant (tertiary-level) student flows and continue the practice of valorising a Western, Anglophone education. We have also sought to represent, however, alternative and emergent macro-flows of international students and have suggested some of the implications of these. Markets *within* Asia, for example, have become far more relevant for understanding macro-scale mobilities. As previously mentioned, the same large-scale data sets and representations do not appear to exist for younger children. We know, from disparate research studies, that many young children do move internationally for education (e.g., Lo et al., 2017), but this has been often framed as a family or household move, as opposed to the presumed more 'individualised' decision-making of the 'older' international student. Here, there would appear to be some room for further research into the macro-level spaces of school-level international mobility.

We then discussed what we have termed (for the sake of simplicity) 'meso' and 'micro' scales of students' experiences. We have considered the role that cities play in ISM; how urban sites serve to attract international students whilst also being transformed by their presence. We considered the function that cities play in creating the 'imaginative geographies' of international students (Kölbel, 2020; Beech, 2014; Lee, 2021; Brooks and Waters, 2018). We also briefly discussed how students contribute to the process of studentification and their daily consumption practices, thereby impacting the urban geographies of particular cities. Some urban areas, such as 'education hubs', are being developed with the principal aim of attracting international students from the surrounding region. In recent years, some research has developed on the spaces of student mobilities that might apply to younger (pre-school and school-age) children also. For example, Leung and Waters (2021) discuss the 'mobility industry' attached to what has been called 'cross border' or 'cross boundary' schooling. Mobility occurs between two cities (in this case, Hong Kong and Shenzhen) and yet across a hard political boundary. These children grow up between two cities and for many their mobility is daily. The city space (and the wider state structures) is crucial to understanding children's educational experiences in this instance (see also Chee, 2017).

Finally, in this chapter we have discussed students' experience at the micro-scale – largely (in the literature) represented by their everyday social experiences and encounters. Here we discussed interactions in the classroom, cafeteria, and on the 'campus' as a whole. We also considered student housing and accommodation. There is still relatively little known about international students' quotidian experiences at the micro-level – extant research tends to focus on international students' 'adaptation' difficulties or struggles, and a broader perspective on a range of experiences (positive as well as negative) would be welcomed. When it comes to younger children, there remains a gap in our knowledge of their experiences at the micro-scale also. As mentioned, children's educational migration tends to be framed in relation to 'the family' and wider household strategies (Waters and Wang, forthcoming) and not viewed as significant in and of itself.

International student mobility is such a profoundly geographical process and yet the depictions of its geographies, within the grey literature, are often limited. The macro-scale (emphasising the international in ISM) paints only a partial picture, ignoring the impacts that students are having on, for example, urban areas and the role of cities in attracting international students and the everyday experiences of international students, as well as largely ignoring the educational mobilities of younger children. During the COVID-19 pandemic, international students' geographies have in many cases contracted dramatically – travel restrictions have meant that their spaces of engagement limited to their student room, a local supermarket and, in some cases, a local food bank. This calls for a revaluation of the *multiple* scales within which ISM is experienced.

More research is needed on the scales and experiences of educational mobilities – both for the 'older' young people (at university) reviewed here and for younger students who often move at the behest of other family members. How, for example, at the macro-scale do states attempt to attract younger students? We know of the impact of the 'Long-Term Social Visitor Pass in Singapore' on the families of young children moving from China with their mothers to study (Huang and Yeoh, 2005). But a wider sense of how other states court (or otherwise) younger mobile children is missing from the literature. Likewise, how do younger international students experience the cities in which they come to live? What are their daily, micro-level interactions, beyond those with family members (which we know relatively more about) (see Yoon et al., 2018)? International student mobility is a vibrant and growing area of academic interest, and viewing it in terms of the macro-, meso-, and micro-scale helps to highlight the interconnectedness of the process as well as the multiplicity of space in relation to children's geographies and education.

Notes

1 It is notable that we do not have comparable data for students at earlier (pre-tertiary) levels of education.
2 We have little knowledge of if and how states engage in attracting younger children at pre-university level. This represents a clear 'gap' in the academic literature. Our sense is, however, that this does not represent the same perceived 'investment' (in human capital) or 'payoff' (with respect to international tuition fees) for states hoping to attract university students from abroad.
3 It is worth noting the recent closure of NUS-Yale College (a partnership between the National University of Singapore and Yale University in the US) (Sharma, 2021) marking, some argue, a failure of this urban hub strategy.

References

Altbach, P. G. (2013) Globalization and forces for change in higher education. In *The International Imperative in Higher Education* (pp. 7–10). London: Brill Sense.
Anderson, J., Sadgrove, J., & Valentine, G. (2012) Consuming campus: Geographies of encounter at a British university. *Social and Cultural Geography*, 13(5), pp. 501–515.
BBC (2021) www.bbc.co.uk/news/uk-england-hampshire-58543260

Beech, S. (2018) Adapting to change in the higher education system: International student mobility as a migration industry. *Journal of Ethnic and Migration Studies*, 44(4), pp. 610–625.

Beech, S. (2019) *The Geographies of International Student Mobility. Spaces, Places and Decision-Making*. Basingstoke: Palgrave.

Beech, S. E. (2014) Why place matters: Imaginative geography and international student mobility. *Area*, 46(2), pp. 170–177

Brooks, R., & Waters, J. (2011) *Student Mobilities, Migration and the Internationalization of Higher Education*. New York: Springer.

Brooks, R., & Waters, J. (2017) *Materialities and Mobilities in Education*. London: Routledge.

Brooks, R., & Waters, J. (2018) Signalling the 'multi-local' university? The place of the city in the growth of London-based satellite campuses, and the implications for social stratification. *Social Sciences*, 7(10), pp. 195.

Brooks, R., & Waters, J. (2022) Partial, hierarchical and stratified space? Understanding 'the international' in studies of international student mobility. *Oxford Review of Education*, pp. 1–18.

Calvo, D. (2018) Understanding international students beyond studentification: a new class of transnational urban consumers. The example of Erasmus students in Lisbon (Portugal). *Urban Studies*, 55(10), pp. 2142–2158.

Chacko, E. (2020) Emerging precarity among international students in Singapore: experiences, understandings and responses. *Journal of Ethnic and Migration Studies*, 1–17.

Chee, W. C. (2017) Trapped in the current of mobilities: China-Hong Kong cross-border families. *Mobilities*, 12(2), pp. 199–212.

Collins, F. (2014) Globalising higher education in and through urban spaces: Higher education projects, international student mobilities and trans-local connections in Seoul. *Asia Pacific Viewpoint*, 55(2), pp. 242–257.

Collins, F. L. (2008) Bridges to learning: International student mobilities, education agencies and inter-personal networks. *Global Networks*, 8(4), pp. 398–417.

Collins, F. L. (2010) International students as urban agents: International education and urban transformation in Auckland, New Zealand. *Geoforum*, 41(6), pp. 940–950.

Collins, F. L. (2012) Researching mobility and emplacement: Examining transience and transnationality in international student lives. *Area*, 44(3), pp. 296–304.

Collins, F. L. (2014) Globalising higher education in and through urban spaces: Higher education projects, international student mobilities and trans-local connections in Seoul. *Asia Pacific Viewpoint*, 55(2), pp. 242–257.

Collins, F. L., & Ho, K. C. (2014) Globalising higher education and cities in Asia and the Pacific. *Asia Pacific Viewpoint*, 55(2), 127–131. doi:10.1111/apv.12050

Fincher, R., & Shaw, K. (2009) The unintended segregation of transnational students in central Melbourne. *Environment and Planning A*, 41(8), pp. 1884–1902.

Fincher, R., & Shaw, K. (2011) Enacting separate social worlds: 'International' and 'local' students in public space in central Melbourne. *Geoforum*, 42(5), pp. 539–549.

Findlay, A. M. (2011) An assessment of supply and demand-side theorizations of international student mobility. *International Migration*, 49(2), pp. 162–190.

Findlay, A. M., King, R., Smith, F. M., Geddes, A., & Skeldon, R. (2012) World class? An investigation of globalisation, difference and international student mobility. *Transactions of the Institute of British Geographers*, 37(1), pp. 118–131.

Finn, K., & Holton, M. (2019) *Everyday Mobile Belonging: Theorising Higher Education Student Mobilities*. Bloomsbury Publishing.

França, T., Alves, E., & Padilla, B. (2018) Portuguese policies fostering international student mobility: A colonial legacy or a new strategy? *Globalisation, Societies and Education*, 16(3), pp. 325–338.

Garmendia, M., Coronado, J. M., & Ureña, J. M. (2012) University students sharing flats: When studentification becomes vertical. *Urban Studies*, 49(12), pp. 2651–2668.

Geddie, K. (2015) Policy mobilities in the race for talent: Competitive state strategies in international student mobility. *Transactions of the Institute of British Geographers*, 40(2), pp. 235–248.

Hall, S. M. (2019) Everyday austerity: Towards relational geographies of family, friendship and intimacy. *Progress in Human Geography*, 43(5), pp. 769–789.

He, S. (2015) Consuming urban living in 'villages in the city': Studentification in Guangzhou, China. *Urban Studies*, 52(15), pp. 2849–2873.

Holton, M. (2016a). The geographies of UK university halls of residence: Examining students' embodiment of social capital. *Children's Geographies*, 14(1), pp. 63–76.

Holton, M. (2016b). Living together in student accommodation: Performances, boundaries and homemaking. *Area*, 48(1), pp. 57–63.

Holton, M., & Riley, M. (2013) Student geographies: Exploring the diverse geographies of students and higher education. *Geography Compass*, 7(1), pp. 61–74.

Hopkins, P. (2011) Towards critical geographies of the university campus: Understanding the contest experiences of Muslim students. *Transactions of the Institute of British Geographers*, 36, pp. 157–169.

Huang, S., & Yeoh, B. S. (2005) Transnational families and their children's education: China's 'study mothers' in Singapore. *Global Networks*, 5(4), pp. 379–400.

Jazeel, T. (2019) *Postcolonialism*. London: Routledge.

Jöns, H., & Hoyler, M. (2013) Global geographies of higher education: The perspective of world university rankings. *Geoforum*, 46, pp. 45–59.

Kenna, T. (2011) Studentification in Ireland? Analysing the impacts of students and student accommodation on Cork City. *Irish Geography*, 44(2–3), pp. 191–213.

Kleibert, J. M. (2022) Transnational spaces of education as infrastructures of im/mobility. *Transactions of the Institute of British Geographers*, 47(1), pp. 92–107.

Knight, J. (2018) International education hubs. In P. Meusburger, M. Heffernan, & L. Suarsana (eds.) *Geographies of the University* (p. 676). New York: Springer Nature.

Koh, S. Y. (2017) *Race, Education and Citizenship. Mobile Malaysians, British Colonial Legacies and a Culture of Migration*. Basingstoke: Palgrave.

Kölbel, A. (2020) Imaginative geographies of international student mobility. *Social & Cultural Geography*, 21(1), pp. 86–104.

Lee, J. (2021) When the world is your oyster: International students in the UK and their aspirations for onward mobility after graduation. *Globalisation, Societies and Education*, 1–14.

Lee, K. H. (2020) Becoming a bona fide cosmopolitan: Unpacking the narratives of Western-situated degree-seeking transnational students in China. *Social & Cultural Geography*, pp. 1–19.

Leung, M. W., & Waters, J. L. (2021) Making ways for 'better education': Placing the Shenzhen-Hong Kong mobility industry. *Urban Studies*, 00420980211042716.

Lo, A., Abelmann, N., Kwon, S. A., & Okazaki, S. (Eds.). (2017) *South Korea's Education Exodus: The Life and Times of Early Study Abroad*. Seattle, WA: University of Washington Press.

Ma, Y. (2020) *Ambitious and Anxious. How Chinese College Students Succeed and Struggle in American Higher Education*. New York: Columbia University Press.

Madge, C., Raghuram, P., & Noxolo, P. (2009) Engaged pedagogy and responsibility: A postcolonial analysis of international students. *Geoforum*, 40(1), pp. 34–45.

Massey, D. (2005) *For Space*. Milton Keynes: The Open University.

Mittelmeier, J., Rienties, B., Tempelaar, D., & Whitelock, D. (2018) Overcoming cross-cultural group work tensions: Mixed student perspectives on the role of social relationships. *Higher Education*, 75(1), pp. 149–166.

Moe (2019) Statistical report on international students in China for 2018. Available at: http://en.moe.gov.cn/news/press_releases/201904/t20190418_378586.html#:~:text=Figures%20show%20that%20in%202018,or%200.62%25%20compared%20to%202017.

Müller, M. (2021) Worlding geography: From linguistic privilege to decolonial anywheres. *Progress in Human Geography*, 0309132520979356.

Nakazawa, T. (2017) Expanding the scope of studentification studies. *Geography Compass*, 11(1), e12300.

Nikula, P. T., & Kivistö, J. (2020) Monitoring of education agents engaged in international student recruitment: Perspectives from agency theory. *Journal of Studies in International Education*, 24(2), pp. 212–231.

OECD (2018) *Education at a Glance*. Paris: OECD.

OECD (2019) *Education at a Glance*. Paris: OECD.

Prazeres, L., Findlay, A., McCollum, D., Sander, N., Musil, E., Krisjane, Z., & Apsite-Berina, E. (2017) Distinctive and comparative places: Alternative narratives of distinction within international student mobility. *Geoforum*, 80, pp. 114–122.

QAA (2014) *London Campuses of UK Universities: Overview report of a thematic enquiry by the Quality Assurance Agency for Higher Education*. Available at: http://dera.ioe.ac.uk/21786/1/London-campuses-of-UK-universities.pdf

Raghuram, P. (2013) Theorising the spaces of student migration. *Population, Space and Place*, 19(2), pp. 138–154.

Raghuram, P., Breines, M., Gunter, A. (2020) Beyond #FeesMustFall: International students, fees and everyday agency in the era of decolonisation. *Geoforum*, 109, pp. 95–105.

Revington, N., & August, M. (2020) Making a market for itself: The emergent financialization of student housing in Canada. *Environment and Planning A: Economy and Space*, 52(5), pp. 856–877.

Riaño, Y., Van Mol, C., & Raghuram, P. (2018) New directions in studying policies of international student mobility and migration. *Globalisation, Societies and Education*, 16(3), pp. 283–294.

Robertson, S. (2011) Student switchers and the regulation of residency: The interface of the individual and Australia's immigration regime. *Population, Space and Place*, 17(1), pp. 103–115.

Rose-Redwood, C. and Rose-Redwood, R. (2019) Self-segregation of global mixing? Social interactions and the international student experience. In C. Rose-Redwood & R. Rose-Redwood (eds.) *International Encounters. Higher Education and the International Student Experience* (pp. 19–34). London: Rowan and Littlefield.

Sassen, S. (1991) *The Global City: New York, London, Tokyo*. Princeton, NJ: Princeton University Press.

Sharma, Y. (2021) Students, faculty angry over closure of NUS-Yale college, *University World News*, September.

Sidhu, R. K. (2006) *Universities and Globalization: To Market, to Market*. New York: Routledge.

Sidhu, R. K., Chong, H. K., & Yeoh, B. S. (2019) *Student mobilities and international education in Asia: Emotional geographies of knowledge spaces*. Springer Nature.

Sidhu, R., Collins, F., Lewis, N., & Yeoh, B. (2016) Governmental assemblages of internationalising universities: Mediating circulation and containment in East Asia. *Environment and Planning A*, 48(8), pp. 1493–1513.

Sidhu, R., Ho, K. C., & Yeoh, B. (2011) Emerging education hubs: The case of Singapore. *Higher Education*, 61(1), pp. 23–40.

Sin, C., Antonowicz, D., & Wiers-Jenssen, J. (2021) Attracting international students to semi-peripheral countries: A comparative study of Norway, Poland and Portugal. *Higher Education Policy*, 34(1), pp. 297–320.

Sin, I. L. (2009) The aspiration for social distinction: Malaysian students in a British university. *Studies in Higher Education*, 34(3), pp. 285–299.

Smith, D. (2005) Patterns and processes of 'studentification' in Leeds. *The Regional Review*, 12, pp. 14–16.

Smith, D. P., & Holt, L. (2007) Studentification and 'apprentice' gentrifiers within Britain's provincial towns and cities: Extending the meaning of gentrification. *Environment and Planning A*, 39(1), pp. 142–161.

Smith, D. P., & Hubbard, P. (2014) The segregation of educated youth and dynamic geographies of studentification. *Area*, 46(1), pp. 92–100.

University World News (2020) www.universityworldnews.com/post.php?story=202006 10154557289

Waters, J., & Z. Wang (forthcoming) Families in educational migration: strategies, relations and emotional investments. In J. Waters & B. Yeoh (eds.) *Handbook of Migration and the Family*. Cheltenham: Edward Elgar

Waters, J. L. (2009) Transnational geographies of academic distinction: The role of social capital in the recognition and evaluation of 'overseas' credentials, *Globalisation, Societies and Education*, 7(2), pp. 113–129.

Waters, J. L. (2017) Education unbound? Enlivening debates with a mobilities perspective on learning. *Progress in Human Geography*, 41(3), pp. 279–298.

Waters, J. L. (2018) International education is political! Exploring the politics of international student mobilities. *Journal of International Students*, 8(3), pp. 1459–1478.

Waters, J., & Brooks, R. (2021) *Student migrants and contemporary educational mobilities*. Cham: Palgrave Macmillan.

Yang, P. (2020) Toward a framework for (re) thinking the ethics and politics of international student mobility. *Journal of Studies in International Education*, 24(5), pp. 518–534.

Yoon, E. S., Lubienski, C., & Lee, J. (2018) The geography of school choice in a city with growing inequality: The case of Vancouver. *Journal of Education Policy*, 33(2), pp. 279–298.

Yuan, W. (2011) Academic and cultural experiences of Chinese students at an American university: A qualitative study. *Intercultural Communication Studies*, 20(1), pp. 141–157.

3

GEOGRAPHIES OF EDUCATION SPACES

Architecture, materialities, power, and identity

Peter Kraftl

Introduction

The environment of an education space is vital to its successful functioning. The 'micro-geographies' of an educational institution – from the layout, décor, and technologies found in classrooms, to the organisation of corridors, playgrounds, and social spaces – all play their part. In turn, the creation of equitable, sustainable, well-maintained learning environments has become a global ambition – enshrined, for instance, in Goal 4a of the UN's Sustainable Development Goals (https://sdgs.un.org/goals). Consequently, the important question of the relationship between education *spaces* and a range of learning outcomes has engaged scholars for several decades (for a review, see UNESCO, 2022). Interestingly, when it comes to the rather narrower question of the direct relationship between education spaces and *learning* (arguably the predominant question for research on school spaces), there is still some uncertainty (e.g., about the impacts of education spaces on learning outcomes) (Baars et al., 2021). However, it is now widely accepted that education spaces – and specifically school buildings – can have both positive and negative effects on the *conditions* that support learning (e.g., socialisation, teacher–learner interaction, motivation, and well-being) (OECD, 2017; Daniels et al., 2019).

Recently, several scholars have argued that we must develop research agendas that are less concerned with apparently linear, instrumental measurements of learning outcomes (Uduku, 2015; Benade, 2017, 2021). It has been shown, for instance, that the so-called innovative – flexible, open-plan – classrooms can improve learning but only when the design is aligned with the values and pedagogic practices driving children's learning (Sailer, 2018). Thus, there should be what Cardellino and Woolner (2020: p. 385) term an 'alignment' between curriculum, school values, pupil behaviours, socialisation, senses of well-being, and more (also Carvalho and Yeoman, 2020). Importantly, then, in addition to learning outcomes, it is

DOI: 10.4324/9781003248538-5

important to understand the wider *experience* of education spaces – particularly for educators and learners – as well as the intentions of architects and others who design them (Tse et al., 2015; Daniels et al., 2019).

It is in the aforementioned contexts that research on the geographies of education – and, specifically geographical research on education spaces – has enabled scholarship to extend beyond questions of the relationship between learning environments and learning outcomes (Holloway et al., 2010; Kraftl et al., 2021; Hammond et al., 2022; Waters and Brooks, 2022). Indeed, a good proportion of this work predates the aforementioned calls to diversify research on learning environments. Although the nature of this relationship remains a key question, many geographers of education have sought to ask how else education spaces *matter* to educational experiences, and, particularly, to the experiences of children who attend school. The aim of this chapter, then, is to examine three key areas of research about education spaces that have extended beyond – but nevertheless complement – that on learning outcomes. First, it considers the role of architecture in attempts to signify and embody the educational intentions of school designers, teachers, and policymakers. Opportunities for including learners in the design process are also briefly considered. Second, and by extension, it zooms in further to consider how material things and 'natures' are enlisted in the use of education spaces. Introducing nonrepresentational[1] and new materialist theorisations of materiality, both of these first two sections highlight how a concern with education spaces is neither static nor ignorant of the everyday lives and practices that happen within them. Rather, geographers' understandings of spaces (or 'spatialities') as lively, ever-changing combinations of the material (such as school architectures or furniture) and the social (such as the everyday happenings in a classroom) are central to seeing education spaces as never-finished phenomena. Third, and exemplifying this observation, the chapter turns to the ways in which power dynamics and formations of identity are played out within education spaces.

Constructing educational spaces: school architectures

A key entry point into research on education spaces for children is an examination of school architectures. In many contexts, the construction of specific, separate institutions denoted as 'schools' followed legislation to make education for children compulsory. In other words, through societies' (and particularly governments') successive efforts to define and regulate childhoods – and to expand those definitions from younger to older children – schools have been key sites at which childhoods have been socially constructed (James and James, 2004). Indeed, it has been repeatedly argued by geographers that the social construction of childhood is very commonly the *spatial* construction of childhood – such that key childhood institutions like schools can tell us not only about the broader societal and educational aspirations of those who built them but also about the ways in which they view(ed) children (Holloway and Valentine, 2000; Kraftl, 2006).

In this light, an important strand of research has considered the intentions of the architects, teachers, and other professionals who have constructed schools. Starting with the antecedents of contemporary school buildings, scholars like Burke and Grosvenor (2008, 2013) have analysed in rich historical detail the ways in which school architectures have evolved over time. Significantly, although their work focuses on England, they demonstrate how knowledges about school architectures may 'travel' – for instance as, in the first half of the twentieth century, British school architects visited North American cities to view cutting-edge school designs (Burke and Grosvenor, 2013). Indeed, the internationalisation of architecture as a discipline meant that contemporaneous school designs in many contexts – including the UK – embodied some of the wider principles of the Modernist movement. Amongst fears about the physical health of children, schools – like homes and entire cities – were being planned not only for learning but for health and well-being, with emphases on light, airy spaces that could promote movement and well-being (Gagen, 2004; Gold, 2013; UNESCO, 2022). This section of the chapter examines in greater detail how – through the use of specific learning materials – education spaces were also tied into projects to build and reinforce national identities during the first half of the twentieth century.

Beyond the potential implications for health, historical analyses of school architectures also demonstrate how school buildings are represented in (policy) discourses about education, childhood, and culture. For instance, Uduku's (2018) work has shown how, in colonial Africa, schools were designed according to Christian principles by missionaries until guidance was standardised in the 1940s by colonial governments. After the 1940s, international organisations – particularly UNESCO – became increasingly involved in the articulation of architectural standards and guides for schools that reflected (again) the generation and circulation of international(ising) knowledges about education. As Uduku (2018) shows, UNESCO emphasised the importance of local materials, towards the integration of (Global North) understandings of 'sustainability' into both design and pedagogy, whilst moves for 'child-centred learning' that were influencing the design of open-plan classrooms in the UK in the 1970s (Burke and Grosvenor, 2008) were also replicated in guides for countries in the Global South.

From a critical perspective, the work of Uduku and others is vital for understanding the construction of school buildings as part of wider and multifaceted forms of colonisation. Although one assumes with good intentions, perhaps most powerful is the way in which the imperatives guiding school architectures – past and present – typify the colonisation not only of places but also of childhoods. In a harrowing account, de Leeuw (2009) exemplifies how the twin colonisation of land and childhoods begat forms of violence that enabled the building of modern Canada. Indigenous children were forcibly removed from their families and communities to attend Indian Residential Schools, whose deliberately imposing, European-style facades were a front for an educational approach that attempted to strip children of all aspects of their own cultures – language, behaviour, and identity.

Moreover, to reinforce that separation, the schools were constructed in locations distant from children's families.

Although the term 'colonisation' must be used with care, and although the principles underpinning contemporary school architectures might not harbour such obvious epistemological or physical violence, it is nevertheless useful – even if as a provocation – to ask whether and how more recently constructed schools similarly attempt to construct, regulate, and control children, albeit in different ways. A key trend, which has been critically analysed in a number of studies around the world, has been the enrolment of school architectures (and increasingly their digital infrastructures) in neoliberal education policies (Gulson and Sellar, 2019). Here, school architectures are viewed as a 'policy instrument' (Wood, 2020). In the UK, for instance, the *Building Schools for the Future* programme – a policy that aimed to rebuild or refurbish each of the 3,500 secondary schools in England – placed school buildings centrally within educational policies that harboured the twin goals of increased social inclusion and the promotion of the nation's economic competitiveness on the global stage (Tse et al., 2015). Schools were designed to impress: with large, airy atria and bold designs that were intended to 'inspire' children to attend school and encourage communities surrounding them (and potential investors) to see schools as anchor points for regeneration (Kraftl, 2012). Meanwhile, a focus on design quality – even in terms of small technical and technological details – was intended to reinforce a sense in which the then-government was trying to build a 'new generation' of schools that would breed a new generation of flexible, technologically savvy, responsible learners (and future workers) (Kraftl, 2012; Amsler and Facer, 2017).

The latter reference to 'responsible' learners – and the responsibilisation of children and young people as future citizens of a neoliberal world – has also worked through in school building policies in several contexts. On the one hand, and untethered for a moment from neoliberal imperatives, childhood and education scholars have, for several decades been critically evaluating approaches to children's participation in decision-making around issues (and spaces) that matter to them (Khan et al., 2020). Hence, there is an important and steadily growing body of work that examines the conditions, processes, constraints, and outcomes for children's participation in school design. This often emphasises creative, dialogic, child-led techniques that enable direct interaction between children and architects, which are properly and realistically contextualised in terms of likely outcomes (Birch et al., 2017). As Green (2014) demonstrated in a study based in Australia, not only can the design itself improve, but children may acquire many kinds of skills and knowledges – including around sustainability and connectedness to local places – through the process of participating in design.

On the other hand, there must also be critical reflection on the limits of children's participation (Percy-Smith, 2010). This applies particularly to questions surrounding the political imperatives that may drive participatory agendas. Is, for instance, children's participation used to (further) devolve responsibility for policy successes (or failures) to the population at large? Is it a further tool – and a highly visible and symbolic one when it comes to school architectures – in a wider strategy to encourage children to become self-governing, individuated, neoliberal subjects? These questions exemplify two of the key critical interventions that scholarship on the construction of educational spaces has sought to make: first, in terms of an interrogation of the role of architectural designs, policies, processes, and intentions in articulating and supporting the goals of education; second, in terms of the social and spatial construction of childhood, as those childhood ideals are embedded within the (inter)national demands of particular historical and geographical milieu.

Material and 'natural' features of education spaces

It is hard to determine where the architectural elements of an educational space begin and end – not least when it comes to the many ways in which inside and outside learning environments are decorated, furnished, maintained, improved, and used over time. Although the scalar logic is not a neat one – especially considering how some infrastructures and technologies extend beyond individual buildings or campuses – it is nonetheless useful to attempt to 'zoom in' on some of the material and 'natural' constituents of educational spaces. Whilst the category of the material in itself might be contested, the term 'natural' is perhaps more so (hence the scare quotes), as it speaks of key philosophical and pedagogical differences in terms of how children are conceived in relation to the more-than-human (see Kraftl, 2020, for an overview).

Before turning to debates about 'natures', however, it is again instructive to return to scholarship on the historical geographies of education spaces. Focusing again on the first half of the twentieth century, a number of geographers have examined the inclusion in curricula – including for geography – of models and other material objects. Anticipating a move to child-centred education from the second half of the century onwards, the models – of earth processes – were intended to stimulate hand–eye coordination and hence the physical and mental skills required to 'think geographically' (Ploszajska, 1996). Interestingly, however, those models and other learning materials were simultaneously tied – as with school architectures – to efforts to support children to become active local, national, and global citizens (Matless, 1996). A similar deployment of 'banal' objects in the classroom, in the service of expressions of nationalism, has been documented elsewhere – for instance, through 'overt and creative expressions of nationalism' in textbooks and other teaching aids in Argentina, particularly in light of the ongoing contestation around the Falkland Islands (Benwell, 2014: p. 51).

If the aforementioned historical approaches examined the capacities of learning materials to represent particular politico-cultural ideals, a subset of more contemporary-facing work has taken a different approach to studying material objects in education spaces. Bauer's (2015) call to approach geographies of education 'differANTly' is a generative starting point. Bauer explores how theories of materiality such as Actor Network Theory (ANT) and Science and Technology Studies can help explain the agency of material things in learning environments (also Mulcahy, 2017). Certain objects may be seen to 'act' – especially if a piece of technology or furniture breaks unexpectedly, interrupting the flow of a learning event. More generally, material objects exist in complex relationships or assemblages with humans in such a way that education spaces cannot function without them. Taking an example from Nigeria, Bauer (2015) explains how material objects may be crucial to the translation of learning and teaching styles. Bauer shows how, in a task designed to teach children how to measure length with a ruler, a teacher quickly realised that the ruler was not an appropriate instrument for the children he was teaching. Rather, using a knotted piece of rope and some cards, he created a different system for measuring length, which would be more appropriate to the children's own experiences (and was, ultimately, successful). This shows that education spaces, and 'teaching and learning [are] a continuous product of a precarious translation process[es]' (Bauer, 2015: p. 623). As well as highlighting a different way to research material objects in education spaces, these observations about translation loop back to the previous section about school architectures in two ways: first, in reinforcing the contention that education spaces – including buildings and the materials that comprise or inhabit them – are dynamic and never finished; second, in highlighting how the circulation of educational practices and materials – and the (neo)colonial logics that may drive that circulation – may be questioned or subverted through processes of translation.

Somewhat earlier work on the geographies of education also sought to draw out the capacities of material objects within learning environments. Although drawing in part on ANT and similar theories of materiality, nonrepresentational approaches to education spaces also sought to draw out how differently arranged constellations of bodies, materials, spatial layouts, and performances produced particular kinds of atmospheres or 'affects' (Kraftl, 2006). Working in a Steiner Kindergarten in Wales, Kraftl (2006) has shown how teachers sought to combine different sensory stimuli – the soft touch of wooden and cotton furnishings; the smell and taste of apple crumble; the colours of walls; the use of dance and storytelling – to create atmospheres that were meant to be redolent of a particular idea of 'home'. That idea of 'home' was in turn intended to construct ideas and ideals about childhood – in other words that if young children felt 'at home' in this space, they would feel safe, welcome, and better-disposed to learn (also Watkins, 2017). Hence, nonrepresentational approaches to education spaces have sought to look beyond the agency or affordances of material objects themselves, to how combinations of materials, performances, and sensory stimuli may create atmospheres purported to be crucial to learning but that are hard to pin down in words or otherwise (hence 'non'representational; see also Kellock and Sexton, 2018; Robinson, 2018).

A key concern with the notions of 'home' generated at the Steiner Kindergarten is that they may resonate with only a narrow group of children and their families – for instance, along lines of class, ethnicity, and gender – and may hence be and feel exclusionary for some and may not have the desired impacts on learning outcomes. This concern connects with a wider critique that some approaches to materiality in education spaces (and childhood studies more generally) tend to ignore wider questions of identity and power (Petersen, 2018). Whilst these questions are broached in more detail in the next section of this chapter, it is worth briefly flagging that a burgeoning vein of interdisciplinary scholarship has increasingly attended to these questions and their intersection with the materialities of education spaces. Within the Commonworlds Collective, scholars inspired by feminist, new materialist, and posthumanist theories have sought to question the modern, Western binary of 'humans' versus 'nature' (Taylor and Pacini-Ketchabaw, 2018). Emphasising that humans and non-humans are inextricably linked – inhabiting *common worlds* – and that the divisions between them are blurry at best, their work has sought to unsettle assumptions about how children relate to 'nature' and learn about it. The Collective's rich, sometimes ethnographic, work with predominantly young children has narrated encounters between children and companion species, such as rabbits and raccoons. In those encounters, it is sometimes hard to tell what is happening or being learned – often because 'we' humans simply do not 'know'. It is also difficult to discern who is doing what, to whom, and whether humans are taking responsibility for animals or vice versa (Weldemariam, 2020).

Perhaps, most powerful are those analyses that further unsettle preconceptions about those encounters: on the one hand, when children encounter and respond to death, decay, or disease (Taylor and Pacini-Ketchabaw, 2017); on the other hand, where those encounters – for instance, with rabbits introduced onto the Australian continent by European settlers – are entangled in colonial pasts and presents (Taylor, 2020). Whilst material objects may be entrained in questions that extend beyond race and colonialism, the implications of such approaches for decolonial pedagogies are becoming more clearly mapped out (e.g., Nxumalo and Villanueva, 2020). They also reinforce the argument made in the previous section of this chapter that an interest in the micro-geographies of education spaces need not exclude critical attention to how those micro-geographies articulate or complicate educational or social challenges at 'bigger' spatial scales – from questions of environmental futures to the legacies of colonialisms (Pacini-Ketchabaw and Clark, 2016).

Power and identity

Although already implicit in the previous sections of this chapter, a key aspect of scholarship in the geographies of education has been to focus on the learners, educators, and other agents who work and study in education spaces. Indeed, as Holloway et al. (2010) point out, this is where the influence of subdisciplinary children's geographies has perhaps most forcefully been felt within the geographical study of education: by affording children (particularly) a voice about the everyday lives of education spaces. In addition to the kinds of experiences indicated in the previous sections, geographers and others have attended to a rich variety of issues, not least when it comes to school building design (Daniels et al., 2019; see Kraftl et al., 2021, for a broad overview). However, in this section, attention is drawn to two key and enduring themes: power and identity.

In terms of power, a key consideration has been with the kinds of power relations that operate in schools and, especially, in the micro-geographies of the classroom, corridor, or school dining hall. Whilst reflecting wider power relations outside educational institutions – for instance, those between children and adults – the microspatial operations of power within educational spaces are also illuminating in terms of how learning institutions function. Often drawing on the work of French philosopher Michel Foucault, geographers have offered rich empirical insights into power relations and tensions that enable particular frames of discipline and surveillance (Gallagher, 2010). For instance, Pike (2008) demonstrated how, in primary school dining halls, adult staff attempted to regulate both the behaviours and the dietary intake of children through a series of detailed spatial rules, which governed where and when children could sit, move around, and leave the dining hall. However, Pike also offered a striking analysis of how children sought to subvert those rules through small acts of disobedience or play – perhaps by moving around the hall in a different direction, perhaps by sharing food, or perhaps by engaging in playful or humorous acts designed to undermine the authority of adult staff. Critically, as other scholars writing on children, education, and food have shown, these acts of 'cat and mouse' operate as a microcosm of the tensions that exist around 'healthy eating' and obesity and the role of (state) educational institutions, like schools, in intervening in areas that complement but extend beyond curricular matters (Pike and Kelly, 2014; Berggren et al., 2021).

Beyond the kinds of tensions that might exist within the classroom, Foucault-inspired scholarship about surveillance has also critically analysed the exercise of power *by* adults – especially in ways that might either be invisible or incontestable by younger learners (Gallagher, 2010; Thornham and Myers, 2012). In some ways, these resonate with the kind of 'carceral circuitry' that is now wired throughout many contemporary institutional and public spaces – as schools become but one site of surveillance through multiple media and technologies (Gill et al., 2018). On the one hand, in a more literal sense, contemporary school buildings may deliberately – through open-plan design – facilitate surveillance by teachers, which may be further reinforced using closed-circuit television cameras and other similar technologies (Hope, 2013). Indeed, young people may be encouraged to survey and govern their own behaviours through the use of technologies – as Goodyear et al. (2019) show in their study of wearable 'healthy lifestyle' technologies, through which students may adapt their behaviours (or not) to conform with the expectations embedded in smart watches and other wearables. On the other hand, the rising significance of 'data' in schools, and of technologies and platforms designed to collect, monitor and report on such data, affords yet another opportunity for surveillance and for children to be encouraged to regulate their behaviours (Williamson, 2017). As Finn (2016) demonstrates, data may – like the materialities discussed in the previous section – be generative of particular atmospheres or affects that, although ostensibly intangible, may drive the values, actions, and relationships of both learners and educators towards displays of particular kinds of academic 'progress'.

The exercise of power in schools also inevitably inflects questions of identity. Beyond questions of national identity outlined in previous sections of this chapter, and beyond the relationships between learners and children highlighted in this, educational spaces are also key sites in which expressions of identity are negotiated between peers. As Valentine (2000) argued, school corridors and classrooms are spaces in which children and young people articulate and experiment with different narratives of identity (Catling and Pike, 2022). Moreover, such narratives may be riven with challenges and opportunities – from bullying to making friends to acquiring nicknames. These kinds of processes – again occurring at the interstices of learning and educational spaces – are vital to the ways in which children socialise, especially in societies where so much of their time is spent in institutions like schools (Melander Bowden and Gustafson, 2021). Additionally, although these kinds of experiences may be dismissed as 'just' what goes on in schools, or 'just' the playfulness of children and young people, they matter, profoundly, to the experience of growing up, especially in contexts where particular groups might be minoritised. For instance, powerful work on university campuses has shown how minoritised ethnic students negotiate those spaces – from dealing with the 'white gaze' to finding 'counter-spaces' where they might feel more comfortable (Samatar et al., 2021). Once again, this scholarship illuminates how the micro-geographies of education spaces are absolutely central to understanding 'wider' societal processes – from critical questions about healthy eating and obesity to the racisms that continue to permeate many learning institutions.

Conclusion

This chapter has sought to elucidate how and why education spaces matter – both to and beyond the learning outcomes that may or may not be produced by 'good' design. Certainly, although many studies show that school architectures may (if indirectly, and in combination with a range of other factors) support pupils' learning, it is also the case that research in the geographies of education has demonstrated that the social and material spaces of education institutions are important in so many other ways. This chapter has focused on the ways in which geographers of education have examined the micro-geographies of education spaces. On the one hand, those spaces matter intrinsically as they may frame, prompt, or undermine the intentions of architects, teachers, or others who design buildings and equally as they may facilitate a range of power-laden processes that can be very particular to a given institution. On the other hand, education spaces are – despite efforts to make them safe and/or secure – not hermetically sealed from the outside world. Wider policy intentions, expressions of (national) identity, and societal norms may be articulated or heightened within educational spaces or may be resisted or renegotiated.

In the context of rich, burgeoning, intra-disciplinary, and often interdisciplinary scholarship taking place on education spaces, this chapter focused on just three key areas of concern. First, it examined the intentions of architects, teachers, and others who design schools. It noted the importance of an awareness of historical processes and precedents for contemporary school-building programmes and, indeed, there is an ongoing need to couch contemporary micro-geographies of education spaces as much within a wider temporal frame as the wider spatial one identified in the previous paragraph. This observation is particularly important when it comes to the significant task of critically analysing the ways in which colonial and neoliberal practices and politics have imbued (and continue to imbue) the very design and use of school buildings, in many different international contexts. As the chapter has shown through the example of Canadian Indian residential schools, to not heed the violence that can accompany such colonial histories would be profoundly dangerous.

Second, the chapter examined a range of ways in which the 'materialities' of education spaces might matter and, in turn, might be studied. Again, and with an eye to the cultural and or symbolic significance of particular learning materials – like models and textbooks – historical analyses are vital in demonstrating how the micro-geographies of the classroom afford a microcosm of wider aspirations for nationhood. However, material objects – and 'natures' – may also operate and be understood in different terms, which may question a range of long-held assumptions about the ways in which children (as learners) are viewed. Sometimes, greater attention needs simply to be afforded to the ability of material objects – like measuring instruments – to enable acts of translation so that learning can become more contextually accessible and relevant. Alternatively, materials (and 'natures') may be enrolled to create certain kinds of affects or atmospheres.

Finally, and building on the two previous sections, the chapter demonstrated how education spaces are key sites at which power and identity are negotiated. Those negotiations can take place in multifarious sites, involve different constellations of actors, and, increasingly, invoke different technologies. From games of 'cat and mouse' in the classroom, to the increasing technologies of surveillance in schools, to the experiences of minoritised or otherwise marginalised learners in a learning institution, education spaces are also sites at which matters well beyond 'education' come to matter. Taken together, this chapter shows how important it is to focus on the micro-geographies of education spaces, especially (but not only) when the specificities of those sites are placed into broader social, educational, historical, and geographical contexts.

Note

1   Nonrepresentational geographers have attempted to examine how particular aspects of the world may be more complex than, come before, or happen too fast for human cognition and/or 'representation' (for instance, in words). Included within nonrepresentational approaches is an awareness – highlighted by feminist new materialist scholars – that humans are neither the 'centre' of the world and that humans must live their lives with a host of 'more-than-human' others – including animals, plants, minerals, and environmental processes. Both approaches seek to find methods and theories to convey some of these complexities.

References

Amsler, S. and Facer, K. (2017) Contesting anticipatory regimes in education: Exploring alternative educational orientations to the future. *Futures*, 94, pp. 6–14.

Baars, S., Schellings, G.L.M., Krishnamurthy, S., Joore, J.P., den Brok, P.J. and van Wesemael, P.J.V. (2021) A framework for exploration of relationship between the psychosocial and physical learning environment. *Learning Environments Research*, 24, pp. 43–69.

Bauer, I. (2015) Approaching geographies of education differently. *Children's Geographies*, 13(5), pp. 620–627.

Benade, L. (2017) Is the classroom obsolete in the twenty-first century? *Educational Philosophy and Theory*, 49(8), pp. 796–807.

Benade, L. (2021) Theoretical approaches to researching learning spaces. *New Zealand Journal of Educational Studies*, pp. 1–16.

Benwell, M.C. (2014) From the banal to the blatant: Expressions of nationalism in secondary schools in Argentina and the Falkland Islands. *Geoforum*, 52, pp. 51–60.

Berggren, L., Olsson, C., Rönnlund, M. and Waling, M. (2021) Between good intentions and practical constraints: Swedish teachers' perceptions of school lunch. *Cambridge Journal of Education*, 51(2), pp. 247–261.

Birch, J., Parnell, R., Patsarika, M. and Šorn, M. (2017) Participating together: Dialogic space for children and architects in the design process. *Children's Geographies*, 15(2), pp. 224–236.

Burke, C. and Grosvenor, I. (2008) *School*. London: Reaktion Books.

Burke, C. and Grosvenor, I. (2013) An exploration of the writing and reading of a life: The "body parts" of the Victorian School Architect ER Robson. In *Rethinking the History of Education* (pp. 201–220). New York: Palgrave Macmillan.

Cardellino, P. and Woolner, P., (2020) Designing for transformation – A case study of open learning spaces and educational change. *Pedagogy, Culture & Society*, 28(3), pp. 383–402.

Carvalho, L. and Yeoman, P. (2021) Performativity of materials in learning: The learning-whole in action. *Journal of New Approaches in Educational Research*, 10(1), pp. 28–42.

Catling, S. and Pike, S. (2022) Becoming acquainted: Aspects of diversity in younger children's geographies. In L. Hammond, M. Biddulph, S. Catling and J.H. McKendrick (eds.) *Children, Education and Geography: Rethinking Intersections* Abingdon: Routledge.

Daniels, H., Stables, A., Tse, H.M. and Cox, S. (2019) *School Design Matters: How School Design Relates to the Practice and Experience of Schooling*. New York: Routledge.

De Leeuw, S. (2009) 'If anything is to be done with the Indian, we must catch him very young': colonial constructions of Aboriginal children and the geographies of Indian residential schooling in British Columbia, Canada. *Children's Geographies*, 7(2), pp. 123–140.

Finn, M. (2016) Atmospheres of progress in a data-based school. *Cultural Geographies*, 23(1), pp. 29–49.

Gagen, E.A., (2004) Making America flesh: physicality and nationhood in early twentieth-century physical education reform. *Cultural Geographies*, 11(4), pp. 417–442.

Gallagher, M. (2010) Are schools panoptic? *Surveillance & Society*, 7(3/4), pp. 262–272.

Gill, N., Conlon, D., Moran, D. and Burridge, A. (2018) Carceral circuitry: New directions in carceral geography. *Progress in Human Geography*, 42(2), pp. 183–204.

Gold, J.R., (2013) *The Experience of Modernism: Modern Architects and the Future City, 1928–53*. New York: Taylor & Francis.

Goodyear, V.A., Kerner, C. and Quennerstedt, M., (2019) Young people's uses of wearable healthy lifestyle technologies; surveillance, self-surveillance and resistance. *Sport, Education and Society*, 24(3), pp. 212–225.

Green, M., (2014) Transformational design literacies: Children as active place-makers. *Children's Geographies*, 12(2), pp. 189–204.

Gulson, K.N. and Sellar, S., (2019) Emerging data infrastructures and the new topologies of education policy. *Environment and Planning D: Society and Space*, 37(2), pp. 350–366.

Hammond, L., Biddulph, M., Catling, S. and McKendrick, J.H. (2022) The child and their (geographical) education. In L. Hammond, M. Biddulph, S. Catling and J.H. McKendrick (eds.) *Children, Education and Geography: Rethinking Intersections* Abingdon: Routledge.

Holloway, S.L. and Valentine, G., (2000) Spatiality and the new social studies of childhood. *Sociology*, 34(4), pp. 763–783.

Holloway, S.L., Hubbard, P., Jöns, H. and Pimlott-Wilson, H., (2010) Geographies of education and the significance of children, youth and families. *Progress in Human Geography*, 34(5), pp. 583–600.

Hope, A. (2013) Foucault, panopticism and school surveillance research. In *Social Theory and Education Research* (pp. 47–63). Abingdon: Routledge.

James, A. and James, A. (2004) *Constructing Childhood: Theory, Policy and Social Practice*. London: Bloomsbury.

Kellock, A. and Sexton, J., (2018) Whose space is it anyway? Learning about space to make space to learn. *Children's Geographies*, 16(2), pp. 115–127.

Khan, M., Bell, S. and Wood, J. (eds.). (2020) *Place, Pedagogy and Play: Participation, Design and Research with Children*. Abingdon: Routledge.

Kraftl, P. (2006) Building an idea: The material construction of an ideal childhood. *Transactions of the Institute of British Geographers*, 31(4), pp. 488–504.

Kraftl, P. (2012) Utopian promise or burdensome responsibility? A critical analysis of the UK government's building schools for the future policy. *Antipode*, 44(3), pp. 847–870.

Kraftl, P. (2020) *After Childhood: Re-thinking Environment, Materiality and Media in Children's Lives*. Abingdon: Routledge.

Kraftl, P., Andrews, W., Beech, S., Ceresa, G., Holloway, S., Johnson, V. and White, C. (2021) Geographies of education: A journey. *Area*. DOI: 10.1111/area.12698

Matless, D. (1996) Visual culture and geographical citizenship: England in the 1940s. *Journal of Historical Geography*, 22(4), pp. 424–439.

Melander Bowden, H. and Gustafson, K., (2021) Embodied spatial learning in the mobile preschool: The socio-spatial organization of meals as interactional achievement. *Children's Geographies*, DOI: 10.1080/14733285.2021.1936454

Mulcahy, D., (2017) The salience of liminal spaces of learning: Assembling affects, bodies and objects at the museum. *Geographica Helvetica*, 72(1), pp. 109–118.

Nxumalo, F. and Villanueva, M.T., (2020) (Re) storying water: Decolonial pedagogies of relational affect with young children. In *Mapping the Affective Turn in Education* (pp. 209–228). New York: Routledge.

OECD (2017) *The OECD Handbook for Innovative Learning Environments*. Paris: OECD.

Pacini-Ketchabaw, V. and Clark, V., (2016) Following watery relations in early childhood pedagogies. *Journal of Early Childhood Research*, 14(1), pp. 98–111.

Percy-Smith, B., (2010) Councils, consultations and community: Rethinking the spaces for children and young people's participation. *Children's Geographies*, 8(2), pp. 107–122.

Petersen, E.B., (2018) 'Data found us': A critique of some new materialist tropes in educational research. *Research in Education*, 101(1), pp. 5–16.

Pike, J., (2008) Foucault, space and primary school dining rooms. *Children's Geographies*, 6(4), pp. 413–422.

Pike, J. and Kelly, P., (2014) *The Moral Geographies of Children, Young People and Food: Beyond Jamie's School Dinners*. New York: Springer.

Ploszajska, T. (1996) Constructing the subject: geographical models in English schools, 1870–1944. *Journal of Historical Geography*, 22(4), pp. 388–398.

Robinson, P.A., (2018) Learning spaces in the countryside: University students and the Harper assemblage. *Area*, 50(2), pp. 274–282.

Sailer, K., (2018) *Corridors, Classrooms, Classification – The Impact of School Layout on Pedagogy and Social Behaviours*. Abingdon: Routledge.

Samatar, A., Madriaga, M. and McGrath, L., (2021) No love found: How female students of colour negotiate and repurpose university spaces. *British Journal of Sociology of Education*, pp. 1–16.

Taylor, A., (2020) Countering the conceits of the Anthropos: Scaling down and researching with minor players. *Discourse: Studies in the Cultural Politics of Education*, 41(3), pp. 340–358.

Taylor, A. and Pacini-Ketchabaw, V., (2017) Kids, raccoons, and roos: Awkward encounters and mixed affects. *Children's Geographies*, 15(2), pp. 131–145.

Taylor, A. and Pacini-Ketchabaw, V., (2018) *The Common Worlds of Children and Animals: Relational Ethics for Entangled Lives*. Abingdon: Routledge.

Thornham, H. and Myers, C.A., (2012) Architectures of youth: Visibility, agency and the technological imaginings of young people. *Social & Cultural Geography*, 13(7), pp. 783–800.

Tse, H.M., Learoyd-Smith, S., Stables, A. and Daniels, H. (2015) Continuity and conflict in school design: A case study from Building Schools for the Future. *Intelligent Buildings International*, 7(2–3), pp. 64–82.

Uduku, O., (2015) Designing schools for quality: An international, case study-based review. *International Journal of Educational Development*, 44, pp. 56–64.

Uduku, O. (2018) *Learning Spaces in Africa: Critical Histories to 21st Century Challenges and Change*. New York: Routledge Taylor and Francis.

UNESCO (2022) *International Science-Based Evaluation of Education (ISEE)*. London: UNESCO. https://mgiep.unesco.org/iseea-report-launch

Valentine, G. (2000) Exploring children and young people's narratives of identity. *Geoforum*, 31(2), pp. 257–267.

Waters, J. and Brooks, R. (2022) Geographies of education at macro-, meso- and micro-scales: young people and international student mobility. In L. Hammond, M. Biddulph, S. Catling and J.H. McKendrick (eds.) *Children, Education and Geography: Rethinking Intersections*. Abingdon: Routledge.

Watkins, M., (2017) Can space teach? Theorising pedagogies of social order. *Geographical Research*, 55(1), pp. 80–88.

Weldemariam, K., (2020) 'Becoming-with bees': Generating affect and response-abilities with the dying bees in early childhood education. *Discourse: Studies in the Cultural Politics of Education*, 41(3), pp. 391–406.

Williamson, B., (2017) Learning in the 'platform society': Disassembling an educational data assemblage. *Research in Education*, 98(1), pp. 59–82.

Wood, A., (2020) Built policy: School-building and architecture as policy instrument. *Journal of Education Policy*, 35(4), pp. 465–484.

4

CHILDREN'S GEOGRAPHIES AND SCHOOLS

Beyond the mandated curriculum

John H. McKendrick

Human geography – the easy option for posh kids?

As the season of goodwill approached in 2019, little was shown by his peers to Danny Dorling, custodian of one of the most prestigious chairs in the discipline, the Halford Mackinder Professor of Geography at the University of Oxford. Led by the president for research and higher education at the Royal Geographical Society (with IBG) and the chair of the council of heads of geography, and co-signed by 87 other senior geographers including 57 other departmental heads, a stinging rebuttal to Dorling's claim that geography in higher education in the UK was a soft option for posh kids (Dorling, 2019) was published in the Times Higher Education Supplement (Blunt et al., 2019). Dorling's turn of phrase was unfortunate in an article that implores geography to become a kinder discipline, but the criticisms that he makes are ones that cannot be ignored. Dorling's concerns are fivefold: he observes that the discipline was rooted in a tradition of generating useful knowledge for those who were expected to administer the British Empire's affairs; he asserts that British geography celebrates that many of its students progress to careers that sustain the prevailing economic system; he is uncharitable on the competencies of many of its students, particularly those from privileged backgrounds; he argues that geography became viewed as a palatable university degree option for those from privileged backgrounds who had not 'done that well' at school; and he notes that British geography departments have a higher-than-average proportion of students from richer homes. He implores UK geography in higher education to redress the imbalance of who we teach and implies that the human geography that we teach might need to be recalibrated for this to be achieved. Blunt et al.'s case for the defence revolved around assertions that a geographical education readied students to tackle complex problems such as poverty, that geography graduates were tackling poverty in their careers, that geography is not a 'soft option', and that the

DOI: 10.4324/9781003248538-6

proportion of disadvantaged students studying school geography had doubled in less than a decade. The examination of British geography's social complexion has continued beyond this initial exchange (Dorling, 2020; Brace and Souch, 2020).

Depending on how the reader chooses to position this chapter, it might be viewed as taking up Dorling's challenge to think more critically about who we teach and what we teach. On the other hand, it might also be viewed as exemplifying what Blunt et al. argue is the utility of geography to address complex issues such as poverty. Undoubtedly, it is an extension of their exchange, as the focus in this chapter is squarely on school geography. I argue that geography has the power to improve the quality of children's lives as lived and that we should seek to realise that potential in the interests of promoting social justice, specifically with regard to tackling child poverty and overcoming problems associated with area deprivation. This will only be achieved when one of the central premises of this book is realised: when we seek a purposively progressive fusion of children's geographies, geographies of education, and geography education. If we believe in the power of geographical thinking, then we also need to think much more seriously about who we teach, what geography we teach in schools, and how we teach those geographies and why.

I start by rethinking how we define purpose in the discipline. As I adopt a case study approach, I provide some essential background to Scotland, and its school geography, before presenting the case for disruptively progressive school geography by reflecting on the scale of geographical education, the character of learning communities, what we teach, and how we teach it. Although focused on Scotland, I argue that this geographical agenda has relevance far beyond.[1]

Refocusing the purpose of our valuable geographical education in schools

No reader of this book needs convincing of the (potential) value of geography. Overt specifications of purpose are outlined by learned geographical societies, university geography departments, and school qualification frameworks, among others. In Scotland, the Scottish Qualification Agency[2] explains that

> [t]he study of geography introduces candidates to our changing world, its human interactions and physical processes. Candidates develop the knowledge and skills to enable them to contribute to their local communities and wider society. The study of geography fosters positive life-long attitudes of environmental stewardship, sustainability and global citizenship. Practical activities, including fieldwork, provide opportunities for candidates to interact with their environment.
>
> *(SQA: n.d.)*

School geography in what is known as the senior phase (ages 15–18 in Secondary Year 4 through Secondary Year 6) in Scotland is structured into content on

(i) physical environments, (ii) human environments, and (iii) global issues and geographical skills.

The focus on global citizenship and global issues might be viewed as a continuation of geography's traditional concern to broaden horizons and enable learners to better understand the nature of the wider world and their place within it. Less charitably, such a focus in Scotland (as with the rest of the UK) might be viewed as a continuation of the narrow Empire mindset that it is our British birthright to assume responsibility for the wider world (Catling, 2014). What is more certain is that all our actions have consequences. Although geography's focus on global issues might draw attention to the ways in which our minority world creates many global problems, it also reinforces the mindset that things are not so bad 'back home' and others (beyond Scotland and the rest of the UK) are much worse off than us. Relatively, this might be so. However, if school geography is blind to the problems that are endured by its own pupils, then it is complicit in their presence, perpetuation, and proliferation. At the very least, we need to strengthen our focus on the geographies of injustice in Scotland, and preferably to fashion a geography that is *for* and *with* disadvantaged pupils.

We might argue that the subject matter of geography is merely a means to an end if our goal is to demonstrate the value of a geographical education and furnish learners with geographical skills or to promote 'thinking geographically'. On the contrary, the subject matter of geography is critically important if we aspire to deliver a purposeful geography that tackles social injustice (which would also demonstrate the value of geographical approaches and furnish learners with geographical skills). Fashioning such an agenda is not without precedent in geography in higher education, most notably with the emergence of the welfare tradition in the 1970s (Smith, 1974). Similarly, much contemporary research in the subdiscipline children's geographies aims to better understand children's everyday lives and to 'centre' the child in this analysis (Holloway, 2014). However, these progressive geographies have a limited reach, perpetuated by reward structures in higher education that prioritise the production of academic knowledge over impactful and transformative outcomes that arise from it. In this chapter, I speculate on the potential for a geographical education that seeks to improve children's lives as lived, focusing on what geography can do to tackle the poverty that is experienced by many in school classrooms.

Case study: Scotland and its geography

Scotland has its own education system, which differs in important ways from the other nations in the UK. School education in Scotland (and Wales) remains more strongly welded to comprehensive education, having not embraced the Academy system that is being promoted in England (Eyles and Machin, 2019) and with almost all state-funded schools under the control of one of Scotland's 32 local authorities. School education is broadly based with pupils who are considered to be the most academically able expected to present for five Scottish Highers (compared to three A Levels elsewhere in the UK). University honours degrees tend to be four years (compared to three elsewhere in the UK).

Since 2010, Scotland's schools have followed the *Curriculum for Excellence (CfE)*, which outlines the national curriculum for nursery, primary, and secondary schools.[3] The purpose of *CfE* is to enable each child or young person to become a successful learner, a confident individual, a responsible citizen, and an effective contributor. A broad general education (for pupils aged 3 to 13/14, straddling nursery education, primary school, and the first three years of secondary school) is followed by a senior phase (to the end of school studies – therefore, lasting between one and three years, depending on when pupils leave school). In the broad phase, learning is organised into eight curricular areas, with learning outcomes specified for five stages across these first 11 years of school education. Geography is somewhat hidden and sits awkwardly in the broad general phase of *CfE*, with what we would understand as geography delivered across two curricular areas, that is, science (which includes a focus on biodiversity and interdependence, energy sources and sustainability, processes of the planet, and earth's materials) and social sciences (which includes a progression of outcomes for a range of themes under the umbrella of 'people, place, and environment'). In the senior phase, geography is an elective, which pupils may choose to pursue qualifications.[4] It is also one of the broadening courses for the Scottish Baccalaureate in Science, and one of the core courses for the Scottish Baccalaureate in Social Science.

Although less social mobility through education in Scotland is attained than desired (McKendrick and Sinclair, 2021), it is an aspiration that is supported through Scottish policy interventions, such as closing the poverty-related attainment gap (Scottish Government, 2021a), contextualised admissions to university (Boliver et al., 2017), and the setting of national targets to ensure that a representative proportion of young people from Scotland's 20% most deprived areas progress to higher education by 2030 (Commissioner for Fair Access, 2021). Furthermore, as with Wales and Northern Ireland, Scotland continues to offer an educational maintenance allowance to support low-income pupils to continue their education beyond the years of compulsory schooling (Scottish Government, 2021b). Traditionally, this idea of someone without means being able through education to realise their full potential is central to the purpose of education in Scotland. Although not absent elsewhere in the UK, arguably the association with social justice and education is strongest in Scotland.

Scotland also has a national commitment to eradicate child poverty by 2030, enshrined in law through the Child Poverty (Scotland) Act (2017). School-level action is specified as part of work to reduce the 'costs of living', which together with 'income from employment' and 'income from social security and benefits in kind' are identified by the Scottish government as the three drivers of child poverty (Scottish Government, 2018). In *Every Child Every Chance* (Scottish Government, 2018), the first tackling child poverty delivery plan, actions were specified to introduce a minimum level for school clothing grants and provide 'further' support to reduce the hidden costs of the school day, with the child poverty measurement framework to include evidence on the school leaver attainment gap, take up of free school meals, and level of school clothing grant. The importance of this work was reaffirmed in *Best Start, Bright Futures*, its second child poverty delivery plan for 2022–2026 (Scottish Government, 2022a).

Potentially powerful geography

Nine of Scotland's 16 non-specialist higher education institutions offer degrees in geography. According to the University and Colleges Admission Service for the UK, 305 applicants were accepted to a geography degree course at university in Scotland in 2021, with 150 specialising in human geography, 100 in physical geographical sciences, and 55 registered for general geography (UCAS, 2021). However, this greatly underestimates the number of undergraduate students who encounter geography in Scottish universities, as the first year of university study in Scotland typically comprises four or five electives, in addition to a foundational module/s in their chosen specialism. Many non-geographers will elect to study geography in the first year of their university degree.

In the senior phase in schools, in any given year, over 10,000 pupils choose geography as one of their (typically) seven National 5 subjects, with over 7,000 progressing to higher studies and almost 800 to advanced higher studies (Scottish Qualifications Agency, 2021). Geography remains a popular option in the senior phase, with the numbers presenting for examinations increasing for all levels between 2020 and 2021. It is the tenth most studied discipline at National 5 level and the eleventh most studied at higher level, although its relative ranking in numbers presenting for examinations is slightly lower at all levels compared to a decade earlier.

As with universities, and although substantial, the number of pupils specialising in geography in the senior phase in Scotland, greatly underestimates exposure to geography in its schools. Every year almost 400,000 pupils in primary schools and 170,000 pupils in secondary schools encounter geography through social studies (and arguably science in primary schools) (Scottish Government, 2021c). On the other hand, it should be acknowledged that in this broad general education phase, geography may not occupy a significant amount of teaching time, and in primary schools at least, it is 'hidden' within the curricular areas of science and social studies, and the division of its content across science and social studies tends to mitigate against an integrated approach to geography.

The key point in this barrage of statistics is that geography remains a discipline with potential, given its reach. Understandably, our focus is often on those who choose us – the pupils who elect to study geography in the senior stages of school and those who specialise in geography in higher education. However, if our aim is to utilise the power of geography for the greater good, then there may be traction to be gained by strengthening the focus on geography for social justice within the broad general learning phase in schools and in the first year of university education.

Who do we teach?

Dorling (2019) contended that the traditional goal of geography in the UK is to learn about the wider world so that UKplc might benefit from what it has to offer. Sometimes, this goal is explicit, as Brown (1952) explained when accounting for the work of the Manchester Society of Commercial Geography, but more commonly it is implicit. Where it is assumed that learners are not significantly and personally impacted in relation to subject matter, there is no compelling reason to reflect on the position of learners in relation to it, for example, in relation to rapid population growth in developing countries or problems of the central business district in UK cities. Back in the real world, Dorling is not alone in challenging the assumption of homogeneity among geography students (for example, see Pirbhai-Illich and Martin, 2022, in Chapter 10). Our pupils may already have more connections to our geographies than we think. The problems we discuss in class will have been, and will be, part of many pupils' life experiences. Furthermore, if teachers of geography are to strengthen our focus on local geographies of social justice, it is critical that we understand our learners and learning communities. The life experiences and social capital that learners bring when confronting geographies of poverty, for example, must inform our classroom practice.

It is estimated that 80,000 pre-school-aged children, 100,000 primary-school-aged children, and 60,000 secondary school-aged children are living in poverty in Scotland, close to one in every four children, with approaching one in five considered to be living in severe poverty (Scottish Government, 2021d). Over 50,000 pupils in secondary schools are entitled to, and have registered for, free school meals on account of living in low-income households (Scottish Government, 2020). While it was already understood that children living in poverty are less well prepared and less well-resourced to engage with learning (Sosu and Ellis, 2014), the scale of the problems that poverty presents became much more evident when school learning decanted to home learning environments during COVID-19-induced lockdowns in 2020 and 2021. Several studies highlighted how some children were disadvantaged on account of the lack of support that parents could provide, that digital exclusion was an everyday reality for far more children than was assumed, that many children were reliant on food freely provided in school, and that home environments were not conducive to learning for all (McKinney, 2021; McKinney et al., 2021; Robertson and McHardy, 2021). If geography is to tackle social injustice, it needs to first consider what must be done to counter the disadvantages already faced by too many of its learners.

As every geographer would expect, these disadvantages are unevenly distributed. Although poverty and multiple deprivation is found throughout Scotland, it is predictably more characteristic of some neighbourhoods, geographical settings, and local authorities (McKendrick and Treanor, 2021). Consequently, social complexion in the schools that serve these communities differs markedly, both in the primary and secondary school sectors (Table 4.1). Poverty and area deprivation are commonplace in some schools (Lochend secondary and Bellsbank primary) but

TABLE 4.1 Selected school pupil profiles

School	Stage	Authority	Area	%	% SIMD1	% SIMD2	% SIMD3	% SIMD4	% SIMD5
Lochend Community High	Secondary	Glasgow	Large Urban	59.7	94.8	2.7	*	2.5	0
Holy Cross High	Secondary	South Lanarkshire	Other Urban	15.6	24.0	18.6	14.7	19.8	22.9
Kirkcaldy High	Secondary	Fife	Other Urban	20.8	30.2	38.1	*	*	31.7
Westhill Academy	Secondary	Aberdeenshire	Other Urban	3.2	0	*	*	21.3	78.8
Scotland (average)	*Secondary*	*Scotland*	*N.A.*	*17.0*	*21.6*	*19.4*	*18.7*	*20.3*	*19.9*
Bellsbank	Primary	East Ayrshire	Remote Rural	45.7	100	*	*	0	0
Condorrat	Primary	North Lanarkshire	Other Urban	11.3	19.2	13.5	25.6	22.9	18.9
St John Ogilvie	Primary	West Lothian	Other Urban	17.6	22.1	21.6	12.8	18.2	25.4
Touch	Primary	Fife	Other Urban	26.1	49.3	7.2	5.8	6.5	31.2
Colquhoun Park	Primary	East Dunbartonshire	Large Urban	22.2	40.9	*	3.0	6.4	49.8
Carolside	Primary	East Renfrewshire	Large Urban	1.2	0.7	1.2	0.8	3.4	93.9
Scotland (average)	*Primary*	*Scotland*	*N.A.*	*21.3*	*22.9*	*19.8*	*18.1*	*20.2*	*19.0*

Source: Scottish Government (2020).

Notes: (1) FSM = Free school meals (2) Free school estimated for primary schools are provided for P4 to P7 only, as there is universal provision of free school meals in P1 to P3. (3) SIMD = Scottish Index of Multiple Deprivation, with SIMD 1 representing those living in the 20% most deprived areas in Scotland. (4) * represents where there were between one and five pupils in a category, the asterisk used to assure anonymity for those pupils.

almost absent in others (Westhill Academy and Carolside primary). Some schools have an even distribution across social strata (Kirkc, Condorrat primary and St John Ogilvie primary) while others are polarised with most pupils either from the very most deprived or very least deprived areas (Holy Cross High, Touch Colquhoun primary). Each school demands a bespoke approach to how social justice is approached, both to overcome it as a barrier to learning and to appraise how geographical skills and knowledge can be used to tackle it.

What might we teach?

There are no shortages of non-curricular interest groups arguing that time within the school day should be found to accommodate learning that would facilitate the personal development of pupils. Among the many examples are Money Advice Scotland's delivery of financial education workshops, the Mentors in Violence Prevention's work of the Violence Reduction Unit, Keep Scotland Beautiful's promotion of Eco-Schools, and the Soil Association Scotland's promotion of Food for Life Scotland (campaigning for healthier school meals). While some groups and parents have rejected some content on religious or moral grounds, the main barrier to extending the range of school learning is the finite resource of the time that is available within the school day.

With learning time already under pressure, there is little prospect of extending the amount of geography that is taught within the school day. In Scotland, there is already a very limited amount of time for geography teaching in the broad general phase of education in Scotland, sitting as it does within the curricular area of social studies. Further undermining the prospects for introducing new geographies to children in schools is the reality that the tried-and-tested geographical content would appear to have served pupils well for generations: the metric of the numbers electing to present for qualifications in the senior stage is testament to this. The subfield of children's geographies will not be the only specialist subfield of geography to have emerged in higher education in recent decades that is frustrated that the school geography syllabus cannot be enriched with an infusion of its interests, issues, and approaches to geography.

The way ahead might be to promote the geographical rather than geography. That is, thinking of ways in which geographical skills and a geographical mindset ('thinking geographically') might be brought to bear on all manner of issues that present in schools and that are covered in classroom learning. The proposal is not for geography to colonise the curriculum for the sake of it; rather, the argument is grounded in the belief in the value of geographical approaches, thinking, and skills. More of what we teach in schools would be enriched through geographical analysis – much of which is not currently understood to be 'geography'. It follows that positioning geography as part of social studies need not be a threat to its status. These reflections on the status of geography in schools are not a distraction from the central purpose of this chapter, which is to argue for a school geography that seeks to improve the quality of children's lives as lived. Rather, the geographical analysis that is being promoted would provide additional opportunities for a purposively progressive geography *for* children. From the many ways in which this could be achieved in Scotland, three examples are suggested here – rethinking place, challenging school injustices, and contributing to national priorities.

The social studies curriculum in Scotland could be read as embracing history (people, past events, and societies), geography (people, place, and environment), and modern studies (people in society, economy, and business). Thus, much of what is taught as social studies might not be understood as geography or not presented to pupils as geography. Consider the first row in Table 4.2, which describes recommended outcomes for one of the themes under 'people in society, economy, and business', and the lower row, which describes outcomes for a theme within 'people, past events, and societies'. Although perhaps unlikely to be conceived as geography, these learning outcomes engage subject matter that is at the heart of our discipline, for example, exploring places (SOC 1–02a in Table 4.2), thinking critically about multi-scalar identities (SOC 4–02a), thinking critically about the scale at which social support should be provided (SOC 4–16a), and understanding communities (SOC 2–16a).

The potential afforded by thinking differently about social studies extends beyond finding geography wherever we look. Geography has a rich tradition of profiling places and mapping geographical inequities in access to services. This work is critical in alerting us to the realities of the geographies of poverty and area deprivation. However, there is also risk that area labelling creates its own problems and stigma. Extending our geographical analyses of poverty to embrace issues that seem to 'belong' to other social studies at once enriches our geographical analysis and social studies enquiry. We should not be content to draw attention to the persistence of poverty and area deprivation in areas such as Easterhouse in Glasgow and Raploch in Stirling (McKendrick and Treanor, 2021). We might also want to raise teacher and pupil awareness of those local champions of place who are challenging injustices (such as the work of Cathy McCormack (2009) in tackling housing conditions in Easterhouse in Glasgow), a learning goal of SOC 4–16a in Table 4.2.

Most importantly, there are opportunities to think critically about the extent to which these adult constructions of place adequately reflect the lived reality of children and young people. For example, the so-called deprived areas (such as the Raploch housing estate on the edge of Stirling) are often criticised for not providing ready access to employment, for the lack of community infrastructure, for the poor quality of housing, and the unruly behaviours of its young people. Consideration might be given – a local variant of learning goal SOC 2–02a in Table 4.2 – to exploring local success stories which demonstrate how the so-called deprived areas have provided unique opportunities for some children to thrive (such as the 'rich' environment that allowed Billy Bremner, the professional football player, to hone his skills as a youth in the Raploch – Figure 4.1).

Geography's contribution to promoting social justice is not limited to core curricular: there is a need for geographical skills and knowledge to be deployed to tackle wider injustices that frame school experiences. Is it acceptable that children

TABLE 4.2 Extract for 'Society, Economy and Business' from *Social Studies: Experiences and Outcomes*

	Curriculum for Excellence Levels in the Broad General Stage of Education				
	Early (pre-school and P1)	*First (P2–P4)*	*Second (P5–P7)*	*Third (S1–S3)*	*Fourth (S1–S3)*
	By exploring my local community, I have discovered the different roles people play and how they can help. SOC 0–16a	I can contribute to a discussion of the difference between my needs and wants and those of others around me. SOC 1–16a	I can explain how the needs of a group in my local community are supported. SOC 2–16a I can gather and use information about forms of discrimination against people in societies and consider the impact this has on people's lives. SOC 2–16b I can discuss issues of the diversity of cultures, values, and customs in our society. SOC 2–16c	I can explain why a group I have identified might experience inequality and can suggest ways in which this inequality might be addressed. SOC 3–16a	I can contribute to a discussion on the extent to which people's needs should be met by the state or the individual. SOC 4–16a Through discussion, I have identified aspects of a social issue to investigate and by gathering information I can assess its impact and the attitudes of the people affected. SOC 4–16b I can analyse the factors contributing to the development of a multicultural society and can express an informed view on issues associated with this. SOC 4–16c
	I can make a personal link to the past by exploring items or images connected with important individuals or special events in my life. SOC 0–02a	By exploring places, investigating artefacts, and locating them in time, I have developed an awareness of the ways we remember and preserve Scotland's history. SOC 1–02a	I can interpret historical evidence from a range of periods to help build a picture of Scotland's heritage and my sense of chronology. SOC 2–02a	I can make links between my current and previous studies and show my understanding of how people and events have contributed to the development of the Scottish nation. SOC 3–02a	I have developed a sense of my heritage and identity as a British, European, or global citizen and can present arguments about the importance of respecting the heritage and identity of others. SOC 4–02a

Source: Scottish Qualifications Agency (2009).

Career history

Billy Bremner was born on the 9th of December in 1942 and grew up in the Raploch estate on the outskirts of Stirling. He has a place in both the Scottish and English Football Halls of Fame, following an illustrious playing career with Leeds United, Hull City and Scotland. Billy won 54 caps and scored 3 goals for Scotland. With Leeds United, he played in four European finals (winning the Fairs Cup (early version of the Europa Cup) in 1968, as well as winning the FA Cup in 1972, the League Championship (now Premiership) in 1969 and 1974, the Charity Shield in 1969 and the Football League Cup in 1968. The biography (*Bremner: The legend of Billy Bremner,* by Bernard Bale) was published by Andre Deutsch in 1998. Billy passed away In December 1997.

Childhood days in Raploch

Billy Bremner grew up in Weir Street in the Raploch estate on the outskirts of Stirling. In the words of Bernard Bale, his "... pre-school education was all about football. Before he had even heard of the 'three Rs', he was learning football's 'three Cs' - control, confidence and competitiveness". His competitive spirit was clear at the informal games he played in the local swing park on Sundays. His best friend Issy described how "each team put a shilling in the hat and whoever won the most games used to take all the cash – the money came from returned pop bottles and odd jobs. Billy's favourite expression was, "Are we going for broke?" I don't ever remember losing a game (with Billy in the team) so we did all right out of it." One of the major influences on Billy's early football was his father James. As Billy himself observed, "My father used to remind me very often that things did not just happen, you had to work hard to make them happen. He was a great help to me, I could not have wished for greater support from my parents." Although naturally talented, Billy practiced hard to get the best from his talent and what he lacked in inches in height, he made up for in tenacity. At school and with Gownahill Juniors. he played alongside boys who were older and bigger than he was. Size didn't matter to the diminuitive Bremner, "I soon discovered one thing about playing against lads who were twice as big as me. I had to make up for my lack of height and weight by getting stuck in just that little bit harder." But it was the work ethic, weaned on the Raploch, that Billy himself attributed most to his success. As he said, "Never give anything less than 100 per cent – in everything that you decide to do. If you want to come out on top you will have to try that little bit harder than the next fellow or you will never make it."

FIGURE 4.1 Growing up in Raploch: a vignette of Bill Bremner

in less deprived areas have a wider choice of subjects to study in the senior stages of schooling compared to those serving more deprived communities (Iannelli et al., 2016)? Why do more pupils in remote small towns and remote rural areas fail to attain the expected levels in literacy and numeracy throughout the general phase of schooling in Scotland (Scottish Government, 2021e)? Why do we not know whether extracurricular activities in schools are reinforcing or challenging access to opportunities between more and less disadvantaged pupils? Why are there such dramatic differences in the uptake of free school meals across schools in Scotland (McKendrick et al., 2019)? To what extent are the 'cost of the school day' interventions impactful across Scotland (Blake Stevenston Associates, 2019)? Is it acceptable that there are persistent differences in rates of progression to university across secondary schools in Scotland (Scottish Government, 2022b)? To what extent do all pupils have access to the resources that enable them to thrive in – and beyond – school (Robertson and McHardy, 2021)? To what extent does tutoring reinforce or challenge injustices in school education (Jerrim, 2017)?

Geographical analysis has much to contribute to the issues. If we do not use our geographical understanding and skills to tackle these issues, then we are part of the problem. We should not be content to teach geography to whoever presents in the classroom. We should be concerned to ensure that all have a fair chance of getting to that classroom in the first instance.

Scotland has a national performance framework which aims to measure the extent to which the nation is making progress towards achieving the kind of Scotland the nation aspires to become (Scottish Government, 2022c). The Scottish Government has also specified a series of ambitious national commitments to fashion its future, working towards a more social just nation. There are commitments to eradicate child poverty by 2030 (Scottish Government, 2018). There are commitments to tackle the climate emergency (Scottish Government, 2021f). There are commitments to adhere to the principles of the United Nations Convention on the Rights of the Child (Scottish Government, 2021g). These three examples are illustrative, rather than exhaustive. Significantly, each commitment has the potential to enrich children's lives, as lived in the here-and-now. As for the school injustices, there is scope for geographical analysis to enrich our understanding of these issues and for school geography to embrace these issues, demonstrating the necessity and power of geographical analysis.

There is clearly scope for geography in schools to deal more directly with much more of the issues that impact directly on children's lives, highlighting the relevance of geography and the power of geographical analysis.

Lessons from social policy for geographical education? From 'nothing about us' to 'nothing about us, without us, is for us'

There is a growing expectation in social policy in Scotland, that those impacted by a policy should be consulted as it is being developed. This is also consistent with the commitment to uphold the UNCRC and afford children the opportunity to express their opinion on matters that impact on them. As Hart (1992) following Arnstein (1969) outlined, there is always a risk of tokenism, with what presents as consultation delivering no more than an illusion of engagement. However, there is firm belief, if not a plethora of irrefutable evidence, that meaningful participation leads to better outcomes.

Each local authority in Scotland must prepare an annual Local Child Poverty Action Report (LCPAR) to outline what is being undertaken to tackle child poverty locally. In 2019, the Poverty and Inequality Commission (2019) reported that it has been asked by the Cabinet Secretary for Communities and Local Government to review the 32 Local Child Poverty Action Reports to understand if people with direct lived experience of poverty had been asked for their views and experiences and to appraise the impact of this engagement on the development of the local action plans. The Commission was disappointed to find that most of the reports it reviewed did not even mention such engagement: on the other hand, it highlighted two reports in which involving those with lived experience was 'making a real difference' to policy and practice. Although focused on child poverty, no mention was made of whether it was children with lived experience who were consulted.

There are lessons to be learned for classroom geography if the progressive potential of children's geographies is to be fully realised. What is sought is more than a geography of children, in which there is a shift from children being an absence to becoming an absent presence in geographical education (see McKendrick, 2001 for parallels in population geography). Rather, what is desired is learning in which school geography affords children meaningful opportunity to understand and confront the social injustices that frame their lives.

Conclusion: a purposively progressive fusion of children's geographies, geography of education, and geography education

Geography in UK universities continues to be populated by a disproportionate share of students from more affluent backgrounds. Unlike Dan Dorling, my experience of these 'posh kids' is not so damning. One of the memories that endures from my time as a lecturer in geography at the University of Manchester in the early 1990s is walking the streets of Bayeux at lunchtime at the end of a Normandy field trip without a penny in my pocket, with an extra helping of an unwanted packed lunch in my backpack, and encountering a group of jovial students devouring bowls of *moules-frites* washed down with wine at a pavement restaurant. They were not at fault for their backgrounds. They were, overall, motivated students who shared a passion for geography, delivered what was required, and were a pleasure to teach. However, I do recall one parent exclaiming 'why are you studying that?' and being perturbed having viewed a research poster on the geography of deprivation, which was on display outside the annual graduation reception. Thus, although I am not unduly concerned with the competency of who we teach in UK higher education, like Dorling (2019), I recognise that there are consequences for what geography can achieve that follow from who we teach and what we teach. Geography can do better.

I believe in the utility of geography, and I believe that education could and should be used to promote social justice – my particular interest is reducing poverty and area deprivation and the problems associated with them. In this chapter, I have argued that we could be more impactful working within the existing curriculum, but that much more could be achieved if we recalibrated our purpose as geographers beyond the mandated curriculum. I am not alone in arguing the point. Back in 2014, Simon Catling outlined principles to give voice to younger children in primary geography, premising his argument on the point that education is political and it is a pretence that school subjects are neutral. Among the principles for an 'empowering pedagogy' was an agenda for social justice in which:

> [C]hildren are to be encouraged to value and work for equity and just approaches in their lives, their class and school, and more widely. This means tackling problems and issues, and at the least trying to understand them even though they cannot resolve them. It infers quizzing those who have responsible and powerful roles and arguing points with them.
>
> *(Catling, 2014: p. 367)*

More recently, Steve Puttick and colleagues (2020) have described how six primary schools have collaborated to develop, share, and improve anti-poverty practice in their communities.

What is proposed is a different kind of powerful geographical knowledge from that which is debated in geography education. There is merit in Young's (2008) conception of powerful knowledge (Lambert, 2011; Maude, 2016); notably the aspiration to give children power over their own knowledge and to provide knowledge that facilitates participation. However, there are limitations and dangers (Rudolph et al., 2018). Although geography celebrates how it broadens horizons, advocates of 'powerful geographical knowledge' assert that geography must take young people beyond their immediate and constrained lives so that the power of this geography can be realised (to develop young people's intellectual powers). Our powerful knowledge needs to be less elitist. As Catling (2014) has shown, we can be powerful by respecting this local knowledge and working alongside it, to better understand and challenge the everyday injustices that too many of our young people endure.

Notes

1   There is no 'United Kingdom school geography', with each of the four nations in the UK having discretion to fashion its own. However, the four nations share a common heritage and politics as part of the United Kingdom, which should shape what might be considered a relevant geographical education in schools across the UK.
2   The SQA is currently Scotland's national awarding and accreditation body. It accredits all qualifications in Scotland, with the exception of degrees. It is one of four such national

agencies in the UK alongside Ofqual (England), DCELLS (Wales), and CCEA (Northern Ireland). The Scottish Government announced in March 2022, that a new public body was to be formed to replace the SQA and assume responsibility for developing and awarding qualifications (www.gov.scot/news/new-national-education-bodies/).

3   The *CfE* narrative was 'refreshed' in 2019, to strengthen focus on its core objectives (https://scotlandscurriculum.scot/): this did not alter its structure and objectives. In 2020, the Scottish Government commissioned the Organisation for Economic Co-operation and Development (OECD) to review *CfE*. At the time of writing, the Scottish Government was reflecting on how it would address the recommendations of the OECD report on *Scotland's Curriculum for Excellence: Into the Future*, published in 2021 (OECD, 2021).

4   Typically, pupils in Scotland present for National 5s in Secondary Year 4, Highers in Secondary Year 5, and Advanced Highers in Secondary Year 6. Broadly speaking, GCSEs elsewhere in the UK are equivalent to National 5s in Scotland (which replaced Standard Grades, which in turn had replaced O Grades). Scottish Highers sit between AS and A Levels. Advanced Highers are broadly equivalent to A Levels (although are worth slightly more points for university entry).

References

Arnstein, S.R. (1969) A ladder of citizen participation. *Journal of the American Institute of Planners*, 35(4), 216–224.

Blake Stevenston Associates (2019) *Evaluation of the Cost of the School Day Programme (2018–19)*. Edinburgh: NHS Health Scotland. www.healthscotland.scot/media/2936/cost-of-the-school-day-evaluation-full-report.pdf

Blunt, A., Evans, M. et al. (2019) Geography degrees are preparing disadvantaged students for relevant careers. *Times Higher Educational Supplement*, 14 December 2019. www.timeshighereducation.com/blog/geography-degrees-are-preparing-disadvantaged-students-relevant-careers

Boliver, V., Gorard, S., Powell, M. and Moreira, T. (2017) *Mapping and Evaluating the use of Contextual Data in Undergraduate Admissions in Scotland. An Impact for Access Project Report*. Edinburgh: Scottish Funding Council. www.sfc.ac.uk/web/FILES/Access/Evaluating_contextual_admissions_Executive_Summary.pdf

Brace, S. and Souch, C. (2020) *Geography of Geography: The Evidence Base*. London: RGS-IBG. www.rgs.org/geography/key-information-about-geography/geographyofgeography/report/geography-of-geography-report-web.pdf/

Brown, T.N.L. (1952) The Manchester society of commercial geography. *The Journal of the Manchester Geographical Society*, 57, 40–45.

Catling, S. (2014) Giving younger children voice in primary geography: Empowering pedagogy – A personal perspective. *International Research in Geographical and Environmental Education*, 23(4), 350–372

Child Poverty (Scotland) Act (2017) (asp 6) [online]. www.legislation.gov.uk/asp/2017/6/contents/enacted

Commissioner for Fair Access (2021) *Re-committing to Fair Access: A Plan for Recovery. Fourth Annual Report*. Edinburgh: Scottish Government. www.gov.scot/publications/re-committing-fair-access-plan-recovery-annual-report-2021/

Dorling, D. (2019) Kindness: A new kind of rigour for British Geographers. *Emotion, Space and Society*, 33, 100630. https://doi.org/10.1016/j.emospa.2019.100630

Dorling, D. (2020) Geography and the Shifting Ratios of Inequality – University, A levels and GCSEs in 2020. *Geography Directions, Blog of the Royal Geographical Society (with IBG)*, 21 August 2020. https://blog.geographydirections.com/2020/08/21/geography-and-the-shifting-ratios-of-inequality-university-a-levels-and-gcses-in-2020/

Eyles, A. and Machin, S. (2019) The introduction of academy schools to England's education. *Journal of the European Economic Association*, 17(4), 1107–1146.

Hart, R.A. (1992) *Children's Participation. From Tokenism to Citizenship. Innocenti Essays, Number 4.* Florence: UNICEF International Child Development Centre.

Holloway, S.L. (2014) Changing children's geographies. *Children's Geographies*, 12(4), 377–392.

Iannelli, C., Smyth, E. and Klein, M. (2016) Curriculum differentiation and social inequality in higher education entry in Scotland and Ireland. *British Educational Research Journal*, 42(4), 561–581.

Jerrim, J. (2017) *Extra Time: Private Tuition and Out-of-School Study, New International Evidence.* London: The Sutton Trust. www.suttontrust.com/wp-content/uploads/2017/09/Extra-time-report_FINAL.pdf

Lambert, D. (2011) Reviewing the case for geography, and the 'knowledge turn'in the English National Curriculum. *The Curriculum Journal*, 22(2), 243–264.

Maude, A. (2016) What might powerful geographical knowledge look like? *Geography*, 101(2), 70–76.

McCormack, C. (2009) *The Wee Yellow Butterfly.* Glendaruel: Argyll Publishing.

McKendrick, J.H. (2001) Coming of age: Rethinking the role of children in population studies. *International Journal of Population Geography*, 7(6), 461–472.

McKendrick, J.H. and Sinclair, S. (2021) Scotland. In Social Mobility Commission (ed.) *State of the Nation 2021: Social Mobility and the Pandemic.* London: SMC. 125–158. https://assets.publishing.service.gov.uk/government/uploads/system/uploads/attachment_data/file/1003977/State_of_the_nation_2021_-_Social_mobility_and_the_pandemic.pdf

McKendrick, J.H. and Treanor, M. (2021) Who lives in poverty? In McKendrick, J.H. et al. (eds.) *Poverty in Scotland 2021: Towards a. 2030 without poverty.* London: CPAG. 85–100. https://askcpag.org.uk/publications/-231001/poverty-in-scotland-2021

McKendrick, J.H., Bouse, D., Connell, D., Ferguson, J., Graham, K., Marshall, K., McRobert, S., McGee, R., Tomassi, L., Vasilev, V., Hughes, T. and Marchbank, J. (2019) *Are Pupils Being Served? A Secondary Review of the Sector's Evidence Base on School Meal Provision at Lunchtime in Scotland.* Glasgow: SPIRU. www.gcu.ac.uk/gsbs/media/gcu/gsbs/SPIRU%20Report%20for%20Assist%20FM%20190826.pdf

McKinney, S. (2021) Poverty and education in the time of the Covid-19 pandemic: Editorial. special edition on poverty. *Researching Education Bulletin. Spring 2021.4-8.* www.sera.ac.uk/wp-content/uploads/sites/13/2021/05/SERA-REB-iss-10-Spring-2021-word-final.pdf

McKinney, S., McKendrick, J.H., Hall, S. and Lowden, K. (2021) What might the Covid pandemic mean for the SERA poverty and education network. *Scottish Educational Review*, 52(2), 4–8. www.scotedreview.org.uk/media/microsites/scottish-educational-review/documents/2020/52-2/McKinney-et-al.pdf

Organisation for Economic Co-operation and Development (2021) *Scotland's Curriculum for Excellence: Into the Future.* Paris: OECD. www.oecd.org/education/scotland-s-curriculum-for-excellence-bf624417-en.htm

Poverty and Inequality Commission (2019) *Poverty and Inequality Commission's Review of the Local Child Poverty Action Reports 2019.* Glasgow: Poverty and Inequality Commission. https://povertyinequality.scot/wp-content/uploads/2019/11/Poverty-and-Inequality-Commission-review-of-the-local-child-poverty-action-reports.pdf

Pirbhai-Illich, F., Martin, F. (2022) De/colonising the (geography) curriculum in Hammond, L. Biddulph, M. Catling, S. McKendrick, J. H. (eds.) *Children, Education and Geography: Rethinking Intersections.* Abingdon: Routledge.

Puttick, S., Hill, Y., Beckley, P., Farrar, E., Luby, A. and Hounslow-Eyre, A. (2020) Liminal spaces constructed by primary schools in predominantly white working-class areas in England. *Ethnography and Education*, 15(2), 137–154.

Robertson, L. and McHardy, F. (2021) *The Poverty-related Attainment Gap: A review of the evidence. A Report for The Robertson Trust.* Glasgow: The Poverty Alliance. www.povertyalliance.org/wp-content/uploads/2021/02/The-Poverty-related-Attainment-Gap-A-Review-of-the-Evidence-2.pdf

Rudolph, S., Sriprakash, A. and Gerrard, J. (2018) Knowledge and racial violence: The shine and shadow of 'powerful knowledge'. *Ethics and Education*, 13(1), 22–38,

Scottish Government (2018) *Every Child, Every Chance. The Tackling Child Poverty Delivery Plan 2018-22.* Edinburgh: Scottish Government. www.gov.scot/binaries/content/documents/govscot/publications/strategy-plan/2018/03/child-chance-tackling-child-poverty-delivery-plan-2018-22/documents/00533606-pdf/00533606-pdf/govscot%3Adocument/00533606.pdf?forceDownload=true

Scottish Government (2020) *Table 3.11 in Pupil Census 2020 Supplementary Statistics. [online].* Edinburgh: Scottish Government. www.gov.scot/publications/pupil-census-supplementary-statistics/

Scottish Government (2021a) *Closing the Poverty-related Attainment Gap: Progress Report 2016 to 2021.* Edinburgh: Scottish Government. www.gov.scot/publications/closing-poverty-related-attainment-gap-report-progress-2016-2021/pages/12/

Scottish Government (2021b) *Education Maintenance Allowances.* Edinburgh: Scottish Government. www.gov.scot/policies/young-people-training-employment/education-maintenance-allowances/

Scottish Government (2021c) *Achievement of Curriculum for Excellence Levels 2020-21. Publication Tables.* Edinburgh: Scottish Government. www.gov.scot/publications/achievement-curriculum-excellence-cfe-levels-2020-21/documents/

Scottish Government (2021d) *Poverty and Child Poverty Data. [online].* Edinburgh: Scottish Government. https://data.gov.scot/poverty/download.html

Scottish Government (2021e) *Summary Statistics for Schools in Scotland. [online].* Edinburgh: Scottish Government. www.gov.scot/publications/summary-statistics-schools-scotland/

Scottish Government (2021f) *Climate Change [online].* Edinburgh: Scottish Government. www.gov.scot/policies/climate-change/

Scottish Government (2021g) *Children's Rights [online].* Edinburgh: Scottish Government. www.gov.scot/policies/human-rights/childrens-rights/

Scottish Government (2022a) *Best Start, Bright Futures. The Second Child Poverty Delivery Plan, 2022-2026.* Edinburgh: Scottish Government. www.gov.scot/publications/best-start-bright-futures-tackling-child-poverty-delivery-plan-2022-26/

Scottish Government (2022b) *School Information Dashboard – Secondary Version [online].* https://public.tableau.com/app/profile/sg.eas.learninganalysis/viz/SchoolInformationDashboard-Secondary/Introduction

Scottish Government (2022c) *National Performance Framework. [online].* https://nationalperformance.gov.scot/

Scottish Qualifications Agency (2009) *Curriculum for Excellence: Social Studies. Experiences and Outcomes.* Livingston: SQA. https://education.gov.scot/Documents/social-studies-eo.pdf

Scottish Qualifications Agency (2021) *Statistics 2021.* [online]. www.sqa.org.uk/sqa/64717.html

Scottish Qualifications Agency (n.d.) *Geography.* [online]. www.sqa.org.uk/sqa/45768.html

Smith, D.M. (1974) Who gets what where, and how: A welfare focus for human geography. *Geography*, 289–297.

Sosu, E. and Ellis, S. (2014) *Closing the Attainment Gap in Scottish Education*. York: Joseph Rowntree Foundation. www.jrf.org.uk/sites/default/files/jrf/migrated/files/education-attainment-scotland-full.pdf

UCAS (2021) *UCAS Analysis and Insights – 2021 Cycle Applicant Figures*. [online]. www.ucas.com/data-and-analysis/undergraduate-statistics-and-reports/ucas-undergraduate-sector-level-end-cycle-data-resources-2021

Young, M. (2008) *Bringing Knowledge Back In*. London: Routledge.

SECTION II

Children's geographies and their significance in, and to, everyday life and education

5

CONNECTING CHILDREN'S AND YOUNG PEOPLE'S GEOGRAPHIES AND GEOGRAPHY EDUCATION

Why this matters to and for children, education, and society

Mary Biddulph, Peter Hopkins, and Simon Tate

Introduction

In this chapter, we consider the intersection between a broader interpretation of the locale of geography education and children and young people's geographies and the mutually beneficial connections that can be made between them. We start with a short discussion focusing on the concept 'thinking geographically' as a means of articulating the purpose of a geography education before considering two cases from research about children's and young people's geographies, the first about ethnic minority children and their negotiations of race and racism and the second about young people's transitions to and through educational contexts. These cases provide valuable insights into the possibilities offered by the intersection between the two fields. We wish to be clear at this stage that we are not here making an argument for framing the school geography curriculum or the whole school curriculum around children's geographies content *per se*. Instead, we are suggesting there is potential for schools and other sites of learning to draw on young people's geographies research in a range of ways to gain new insights into the lives that children and young people experience in and out of school. Such insights have the potential to expose the challenges faced by different groups of children and young people and in so doing serve as an impetus to improve the educational experiences of many. Throughout the chapter we use the phrase 'children and young people' as a means of capturing the wide range of ages reflected in children's geographies research.

Geography education: what does it mean?

Whilst there is no clear-cut answer to this question, there are some opposing viewpoints that are relevant to this chapter. The review of the national curriculum in England, in 2010, marked a shift in policy circles from a more conceptual view

DOI: 10.4324/9781003248538-8

of the curriculum to a strongly content-focused view, sometimes referred to as 'the knowledge turn' in education research (Lambert, 2011). In contrast to this ideological view of knowledge emanating from policy circles in England, many engaged in geography education research internationally (Brooks et al., 2017) agree that the purpose of a school geography education is not to provide young people with what Firth (2013) refers to as an absolutist view of knowledge, namely knowledge that is 'external, fixed, universal and certain' (p. 65) but to develop children and young people's capacity to 'think geographically' (Jackson, 2006), meaning to develop a 'holistic view of the world' (Geographical Association, n.d.) and build a critical understanding of global systems at a range of scales through the content and the concepts that frame the subject. Whilst there is always debate about the precise concepts that frame geography as a discipline (Taylor, 2008), the notion of 'thinking geographically' is helpful because it captures the way a geographical mindset, regardless of the national educational context, can enable young people to use the content and concepts of geography to make connections between the human and the natural world in intelligent and informed ways both in and out of classrooms and understand their own contribution as well as those of others to an ever-evolving world.

Whilst in much of the education literature the notion of 'geography education' refers to the principles and practices of teaching and learning geography, mainly in schools, it is nonetheless important to acknowledge that there are different sites where geographical learning can take place or where an understanding of geographical ideas can support other forms of learning and development. This is exemplified in other curriculum subjects such as history (Boehm et al., 2003), English (language and literature) (Jones and Fitzgerald, 2007), science, technology, engineering, and mathematics (STEM) (Xiaowei et al., 2019) plus more informal school places such as playgrounds, after-school clubs, and school trips (Catling and Pike, 2022). More informal geography education can take place beyond the boundaries of schools also, in homes, local neighbourhoods, shopping streets, play spaces, and other locales accessed and used by children and young people (Kraftl, 2013).

However, developing children's and young people's capacity to develop and apply geographical thinking raises important questions about pedagogical processes in and out of schools. Acquiring the kind of geographical knowledge Firth (2013) describes here requires limited input from children and young people other than to routinely learn and repeat the content as required. However, developing children and young people's capacity to think geographically and apply this thinking to the different spaces they occupy are dependent on pedagogical processes that require the active participation and intellectual engagement of learners. Roberts argues that geographical knowledge can fulfil its potential to develop young peoples' capacity to think geographically only when accessed via what she terms 'powerful pedagogies' (2013a). Drawing on the work of Vygotsky, Bruner, and others, she has long argued for a constructivist epistemology as embedded in the processes and procedures of geographical enquiry (dialogue, debate, analysis, and critique) (2013b). These participatory pedagogies are, she argues, what enable young people to experience for themselves the practices and procedures of knowledge production and critique in a subject such as geography and thus develop their capacity to think geographically.

Roberts (2013a) and Catling and Martin (2011) also contend that young peoples' everyday knowledge – their life experiences in and around the home, their local communities, their immediate and wider social relations – all serve to develop their geographical imaginations that are rich, diverse, and unique and support the development of their geographical thinking. Firth (2018) presents a slightly different perspective arguing that all societies differentiate between esoteric (theoretical) and everyday knowledge and retaining such a distinction is important. However, this is not 'to ascribe a difference in *value*, but rather a difference in *role*' (p. 280).

The following two sections of this chapter present the outcomes of two case studies based on research in children's and young people's geographies. It is important to be clear that each study took place following full ethical approval from relevant institutions and that the young people who participated did so on the basis of giving their own informed consent. Each illustrates something of the breadth of research in children's and young people's geographies, the cross-disciplinary nature of that work, as well as the ways children's geographies intersect with their formal and informal educational experiences: they remind us that despite the formality of a school curriculum, geographical learning and geographical thinking can happen both inside and outside of the formal spaces of school.

Case 1: ethnic and religious minority children and young people: connecting formal spaces of education with wider society

In relation to children's educational experiences, much research has drawn attention to the ways children from ethnic and religious minority backgrounds experience racism within formal educational environments (e.g., Hopkins et al., 2018), including with respect to the nature of the curriculum, the attitudes and values of teachers, and encounters in the school playground. Lander (2015: p. 32) refers to the 'no problem here' discourse that can be reinforced in schools, particularly those with a large white majority. Furthermore, although there have been some efforts to foreground racism in school geography – for example, through the work of the Decolonising Geography group (2021) and the Creative Approaches to Race and In/security in the Caribbean and UK (CARICUK) research project led by Dr Pat Noxolo (CARICUK, 2021). However, teachers, in the main, receive very little formal training about race equality issues and so are often poorly positioned to address racism and racist incidents. Pupils from Roma backgrounds are amongst some of the most excluded in schools (Smith et al., 2020), and a recent enquiry into Islamophobia in Scotland, found increasing levels of anti-Muslim racism in Scottish schools with numerous recommendations made for changes in education, including providing all teachers and lecturers with regular training to counter Islamophobia, integrating an understanding of Islamophobia into compulsory components of the education curriculum, and improving the reporting and recording of Islamophobia in schools (Hopkins, 2021).

Educational environments represent only one context in which ethnic and religious minority children experience racism. These children and young people encounter racism in a many other contexts also not directly associated with their formal education. First, there are informal spaces of education in which ethnic and religious minority children may encounter problematic racialised stereotypes about them or their peers. Second, there is the journey to and from school which can include use of public transport, walking through neighbourhoods with different socioeconomic statuses and varying levels of ethnic diversity. Third, there is negative media coverage about immigration, refugees, and minority groups that filter through into children's everyday lives through TV, social media, and other platforms (Lander, 2015). Fourth, young people may learn racist ideas at home, from family members, family friends, neighbours, and so on. Finally, even if ethnic and religious minority children and young people have not experienced racism directly, they are likely to have close family members who have encountered it and the concerns that this raises for individuals and families often filter through to children and young people.

Using the issue of young people's experiences of racism as an example, we contend that a useful way to think through the connections that can be made between geography education and debates in children's and young people's geographies is to consider the everyday lived experiences of children and young people from diverse ethnic and religious backgrounds. This requires a broad focus that is attentive to the diverse everyday experiences of children and young people, including at home and online, in their neighbourhoods and communities, on their way to and from school, and in their interactions in different public spaces, as well as their engagements in formal educational spaces. Crucially, many of these issues are shaped by political and geopolitical issues such as those related to 9/11 and other terrorist incidents and those associated with political issues such as Brexit (Burrell et al., 2019), whereby specific issues relating to immigration and minorities became increasingly politicised. Efforts to promote anti-racism in school curricula will be more successful if the aim to overcome racism is given a primary focus and if such initiatives are also attentive to the complex ways that racism plays out in the many spaces negotiated by ethnic and religious minority children and young people in their everyday lives (and not only in the classroom).

We now consider two examples from different studies to demonstrate the ways in which connecting geography education with children's and young people's geographies provides an enriched understanding of young people's experiences and negotiations of racism in diverse contexts. The first example draws upon qualitative data from a large study with children and young people from diverse ethnic and religious backgrounds living in urban, suburban, and rural Scotland (Hopkins et al., 2015). In a focus group with young Sikh girls and boys at a suburban secondary school, one of the pupils[1] said:

I remember when I first made like one of my friends like someone like two years ago and she was like, 'What are you?' And I was, 'Well, Sikh'. She was, 'What like a Muslim?' And I was, 'No like Sikh'. And she was, 'Is that not the same as Muslim?' And I was just, 'Oh god, no!' And she was, 'I don't get it, so you are Muslim'. 'No'. People actually just think that if you are brown you are Muslim, and was in school and . . . and this was only like two years ago, she was . . . fifteen. And she didn't even know what a Sikh was. I was just like, oh god! Ha ha. But I think that it is everywhere in Scotland. Obviously how bad it is will change but even in school I think that there is.

This is a clear example of the ways that ethnic minority young people can experience misrecognition at school, in this case, both from their close peers and the wider society. Experiences of misrecognition can be especially damaging as they operate to deny young people's senses of identity and erase their cultural and ethnic background (Hopkins et al., 2017). There were many other accounts of young people becoming increasing tired and worn-down by the need to clarify their identity and explain to others who they are. As this participant explains, this is not just about formal educational spaces but is a challenge 'everywhere in Scotland'.

Our second example draws from five focus groups and ten interviews with young Sikhs in Scotland about the everyday experiences (Hopkins, 2014). In the following quote, a Sikh teenage boy refers to a conversation he had with his father:

I didn't feel scared. I used to wear a white turban a lot when I was about . . . 13 or 14 . . . and my dad said to me, don't wear obviously a white turban, it's sort of what Osama Bin Laden wore at the time and my dad said to me, don't wear a white one. I used to wear blue much more often anyway, but I did occasionally wear a white one and my father did actually prefer me not to do that.

Here, we see that media representations associated with global geopolitics become intertwined with family concerns about the safety and security of their children which leads to conversations about dress choice and bodily presentation. Such decisions are not confined to specific spaces (such as those associated with formal education) but transcend many aspects of the lives of young people including their negotiations of their ethnic and religious identities in public spaces, amongst family and friends, on the way to and from school, and in the school environment.

Case 2: the geographies of student transitions to university

Statistics from the House of Commons' Public Accounts Committee (2018) show that the percentage of students entering higher education from the lowest participation areas of the UK (which strongly correlate with the areas of most social deprivation) is well below the percentage of students entering higher education from the highest participation areas (25%, compared to 59%). This is important as the Public Accounts Committee (2018) also notes that graduates typically earn 42% more than non-graduates. However, equally important to these debates are issues of space and place – both in terms of their impact upon students' decision-making when entering higher education and also in terms of how students subsequently engage with their campuses, neighbourhoods, towns, and cities (Waters and Brooks, 2022). Consequently, as geographers we have an important role to play in exploring the decision-making processes that affect students' transitions to university and to map the social and spatial patterns that these individual decisions can collectively produce.

Of course, cost is an important factor when potential students are deciding whether to attend university; the average student debt is around £50,000 from a three-year degree course (Public Accounts Committee, 2018). However, a House of Commons Briefing Paper in 2018 concluded that there was 'no evidence that those from "lower" socioeconomic groups or (deprived) areas, with historically low levels of participation, have been adversely affected by tuition fees' (Bolton, 2018: p. 13). This begs the question: what other factors are important? We attempted to answer the question by identifying factors that influenced 18-year-old non-traditional students' decision-making about whether and where to study at university (Barnes et al., 2011). Christie (2007) defines non-traditional students as first-generation university attendees from working-class or minority backgrounds, and we found that within this group there were diverse reasons for choosing to continue to study at university. For some of the students, their decision to attend university had been swayed by factors within their personal geographies: because their parents expected them to go or because they innately felt that it was the taken-for-granted next step for them. These students tended to have high levels of what Bourdieu refers to as cultural capital (Bourdieu, 1986). Others were influenced by community and neighbourhood factors which served as sources of fictive kin (Tierney and Venegas, 2006): for example, inspirational teachers encouraging them to study at university, which in turn meant many of their friends were also going to university.

In contrast, amongst those with lower levels of cultural capital a lack of confidence about taking, what was perceived as, a 'leap into the unknown' was a recurring theme. This group identified as important: peers encouraging them not to apply to an 'elitist' university as they wouldn't 'fit in' and teachers reinforcing problematic stereotypes about particular Russell Group universities, which had a negative impact upon students' self-confidence and self-esteem (Barnes et al., 2011). Here, the impact of what Herbert and Thomas (1998: p. 202) refer to as the 'neighbourhood effect' can be seen:

> In addition to the influence of parents and home, children are affected by the values and codes of behaviour prevalent in the residential area in which they live. Neighbourhood values are likely to reinforce home values and geographers have been interested in isolating a neighbourhood effect in education.

One consequence is that non-traditional students tend to cluster in the post-1992 universities, rather than in the Russell Group, creating a socio-spatial divide. Indeed, Reay et al. (2001) argue that a socio-spatial hierarchy has emerged where the top tier of research-led institutions has an overwhelmingly white, middle-class student base. Adding another layer onto this, studies carried out in Liverpool (Holdsworth 2006, 2009a) and Edinburgh (Christie 2007) show that many non-traditional students can feel restricted by their social 'immobility' and so end up spatially immobile and attending their local university. Living at home in turn impacts their experience of being a student: 'working-class students were saturated with a localism that was absent from the narratives of more economically privileged students' (Reay et al., 2001: p. 861). Going yet one step further, Holton and Riley (2013) observe that social divisions on campus can force some non-traditional students who live at home to withdraw both socially and academically from university life, 'missing out' on important parts of the student experience because they are physically detached from the campus and other students for large parts of the day and night. In the short term at least, this impacts the way in which these students engage spatially with the campus, neighbourhood, town, or city in which they reside (Tate and Hopkins, 2020).

To be clear, this body of research isn't elitist – it isn't suggesting that all students should aspire to attend Russell Group universities. Instead, it is seeking to highlight the problems created by the socio-spatial segregation of students at different universities or on the same campus, one of which can be non-traditional students' frustration at their inability to find a way to embody the normative student experience to which they aspire. More positively, our research found that this social and spatial immobility is often temporal and many non-traditional students eventually find their own ways to integrate once at university (Barnes et al., 2011). Some become friends with students from similar socioeconomic backgrounds; some become friends with students from very different backgrounds that they otherwise would not have had the opportunity to meet. Geography students seem to have a particular advantage in this regard as field courses at university are a particularly useful way of helping students to get to know others in their year group. This points to the potential for higher education to be the catalyst for increased social and cultural capital, leading to increased social mobility.

Using the issue of student transitions, particularly the everyday lived experiences of students from diverse social backgrounds, is a useful way to think through the connections that can be made between geography education and debates in children's and young people's geographies. Also, it helps to shed light upon the common misconceptions of a normative 'student' experience, stereotyped by popular portrayals of student life. For example, Grant (1997: p. 102) notes the university presents 'a particular construction of studenthood which for some students is almost impossible to become'. Exploring this requires us as geographers to rethink stereotypical notions of studenthood (Pötschulat et al., 2021) and to broaden our geographical focus beyond the campus to include other spaces (such as the home, neighbourhood, community, and online spaces) – and the relationship between these and the ways in which students interact with public spaces in their home town or city (Chatterton, 2010). The challenge is also to explore these issues in a more inclusive way. With a handful of notable (and somewhat dated) exceptions, overlooked within most of the work on transitions is the way in which geography students understand and negotiate the transition to university (see, as notable exceptions, Barnes et al., 2011; Bryson, 1997; Haigh and Kilmartin, 1999; Maguire et al., 2001; Marriott 2007; Holdsworth, 2006; 2009a, 2009b; Tate and Swords, 2013). Our view is that only if we empower students to speak much more loudly in transitions debates and discourses can we answer questions such as: did students perceive the new subject knowledge required at university (Hammond and Healy, 2022) to be the most problematic aspect of the transition they faced as new undergraduates? Is the 'knowledge turn' as embodied in the revised advanced level examinations for geography (examinations in England, Wales fand Northern Ireland for young people who are generally aged 18 years and above) (see ALCAB, 2014) helping to improve their experience of making the transition to university? What do students think is the best way to improve transitional issues within geography? The answers are complicated, spatial, multi-scalar, and intersectional. Therefore, as Taylor (2009: p. 651) concludes, the contribution of geographers to these debates going forward will remain 'pivotal' and 'significant'.

Connecting children's geographies and geography education: why this matters to and for children, education, and society

In this section we draw together the earlier discussions to consider ways in which children's and young people's geographies as a field of enquiry can better connect to schools for the benefit of children, education more broadly, and wider society.

Why this matters to and for children (and adults who work with them)

As Cases 1 and 2 illustrate, there are a myriad of ways global events and institutional structures (such as schools and universities) have local and personal/individual consequences for young people and their peers and families. The outcome from such research has the potential to enable *all* children and young people, not just some, to apply their geographical thinking to better understand these consequences and how they differently impact on individuals, friends, neighbours, and communities and support children and young people's understanding of each other, demystifying the notion of difference and challenging social and cultural ignorance (Hopkins, 2010).

The intersection of children's and young people's geographies research and school education also matters to children and young people because it provides insights into young people's lives that can enable adults engaged in the process of educating children and young people, in both the formal context of school and also more informal contexts such as youth clubs and extracurricular groups, to challenge the assumptions they have of young people, not just in relation to race but in relation to other identities too, and so create the space for multiple perspectives and different realities to thrive.

The cases reported here in themselves reveal some of the challenges many children and young people encounter in and out of school and offer some important concepts that could support teachers and other adults to better understand the experiences and feelings of many young people. For example, the concept of 'misrecognition' offers a way into understanding not just that misrecognition happens, often unintentionally, but that the process of being misrecognised has negative consequences for children and young people. The negative consequences are not just the tiresome correction of misrecognition, but that misrecognition negatively affects trust between young people and those who work with them (Korkiamäki and Gilligan, 2020), and it undermines their sense of themselves and the different communities to which they belong. Taylor (1994, cited in Hopkins et al., 2017) argues that misrecognition could be deemed a type of social injustice because it can result in social exclusion and isolation, and so whilst misrecognition is significant to individual children's personal experiences in and out of school, it is also about a dynamic process of exclusion on a wider social scale, including in schools.

Why this matters *to and for education*

Children and young people's geographies research reveals that even very young children have views and opinions that they are more than able to articulate when given the opportunity to do so. The conversations with Sikh children in Scotland reported in Case 1 reveal the potential of young people to reflect on their life experiences, with the participatory practices associated with the research demonstrating the power of providing space for children and young people to discuss their experiences on their own terms and with minimal intervention from adults. For some time now, research in children's and young people's geographies has paid careful attention to the ways in which young people engage with and are equal (but different) partners in research processes – their value is the same, but their role is different. The field acknowledges the importance of ethical research practices for all participants including young people, and this stance equates with the expectations of bodies such as The United Nations Convention on the Rights of the Child (United Nations, 1989) and other international privacy and rights requirements. A useful example here is found in the research protocol co-designed by young people following on from the research outlined in Case 1 (Hopkins et al., 2017), where young people outlined their preferences about how they should be involved in research and what processes they would prefer researchers to follow in the process.

Such practices have much to offer education in terms of democratising school systems but also other institutional systems and organisations working with and for young people by engaging all young people in decision-making processes from an early age. The discourse of young people's participation in institutional processes grew in the mid-to-late noughties and took on a global significance with research in both academic geography and education highlighting the importance of shifting decision-making power structures to better enable young people to contribute to the improvement of institutions such as schools (Fielding and Rudduck, 2002; Thomson and Gunther 2006). Yet some schools, such as in England, seem to have forfeited their democratising potential in favour of increased conformity and compliance (Hammond, 2020) including increasing constraints over how young people are allowed to learn; the 'knowledge-turn' at a curriculum level seems now to coincide with a more autocratic pedagogic turn, with any earlier emphasis on participatory processes and geographical enquiry being supplanted by more of a focus on content 'to be learnt and remembered' (Ofsted, 2021).

Children's and young people's geographies also matter to education more generally because of the reliability of the evidence they report about the different lives of the young. As Case 2 demonstrates, schools have a significant role to play in supporting young people's aspirations, yet these aspirations can be undermined because of teachers' or schools' perceptions of the young people they educate (Johnson et al., 2009). Case 2 reminds us that there is no such thing as a 'normative' student, and this applies at school as well as university, and that much needs to be done to address the issues that cause student immobility at an institutional level – at schools and universities but in society more broadly.

There is potential here too for teachers to enhance their understanding and appreciation of young people's lived experiences of inequalities – including those connected with racism and those associated with children and young people from non-traditional backgrounds – but also with respect to other inequalities and injustices lived through by young people (McKendrick, 2022). Such an understanding will contribute to the enhancement of pedagogical practice and the creation of more politically sensitive and nuanced classroom environments.

Why this matters to and for society

In an era where 'levelling up' and 'social mobility' have been central to the discourse of the UK government, the cases here from England and Scotland serve as reminders that education, including geography education in and out of schools, has a central role to play in the processes of life improvement, regardless of national context. We can see from both young people's experience of race and racism, as well as the range of transitional experiences of young people from home to university, that 'levelling up' to improve the life chances of many disadvantaged young people must necessarily take account of these kinds of experiences if change is to happen.

Together, children's and young people's geographies and education more broadly can make a significant contribution to developing young people's geographical thinking in ways that help them to understand the 'unfair outcomes that result from both social processes and institutional decision-making' (Hopkins, 2020: p. 382) as evidenced in academic geography research. The participatory pedagogies cited earlier can teach children and young people to work with evidence, to question the reliability of sources, and to build informed evidence-based arguments on a range of significant issues such as climate change and environmental justice, distributive justice and resource availability, and relational justice, at a range of scales – local, national, and global. The hope being that the opportunity to engage with these processes whilst of school age leave them well-prepared for the challenges of being participatory and engaged citizens when they are adults.

Conclusion

As a field of study, children's and young people's geographies is well placed to contribute to the betterment of children and young peoples' experiences of geographical education in and out of schools. As the label 'children's geographies' suggests, the field covers a wide range of different and layered understandings of the complexities of children's lives and how these connect to local and global forces at a range of scales. Overall then, there is something potentially emancipating about a children's geographies/school geography/broader education nexus as such an intersection has the potential to build young people's critical understanding of themselves, each other, and the society in which they live and enable young people to understand the ways in which power, authority, and other social processes impact on them now and in the future. The participatory pedagogies of developing geographical thinking reflect something of the methods of children's and young people's geographies. Together there is the potential to utilise both to better engage young people in school processes, the kind of engagement that can then empower them for life beyond school.

Note

1   The names of all participants have been removed to protect their confidentiality.

References

ALCAB (2014) *Report of the ALCAB Panel on Geography*. https://alevelcontent.files.word-press.com/2014/07/alcab-report-of-panel-on-geography-july-2014.pdf (accessed 8th September 2021).

Barnes, L. Buckley, A. Hopkins, P. and Tate, S. (2011) The transition to and through university for non-traditional local students: Some observations for teachers *Teaching Geography* 36(2), pp. 70–71.

Boehm, R., Saxe, D.W. and Rutherford, D.J. (2003) The best of both worlds: Blending history and geography in the K-12 curriculum. In *A Summary of the Time-Space Convergence: A Joint History-Geography Curriculum*. Texas: Gilbert. M. Grosvenor Centre for Geographic Education

Bolton, P. (2018) Tuition Fee Statistics. *House of Commons briefing paper 917*, 19 February 2018. Available at: researchbriefings.files.parliament.uk/documents/SN00917/SN00917.pdf (accessed 17th April 2019).

Bourdieu, P. (1986) The forms of capital. In Richardson, J (ed.) *Handbook of Theory and Research for the Sociology of Education*. Westport, CT: Greenwood.

Brooks, C. Butt, G. and Fargher, M. (2017) *The Power of Geographical Thinking (International Perspectives on Geographical Education)* New York: Springer.

Bryson, J.R. (1997) Breaking through the A level effect: A first-year tutorial in student self-reflection. *Journal of Geography in Higher Education* 21(2), pp. 163–169.

Burrell, K., Hopkins, P., Isakjee, A., Lorne, C., Nagel, C., Finlay, R., Nayak, A., Benwell, M.C., Pande, R., Richardson, M., Botterill, K. and Rogaly, B. (2019) Brexit, race and migration. *Environment and Planning C: Politics and Space* 37(1), pp. 3–40.

CARICUK (2021) *About.* https://caricuk.co.uk/about-caricuk/ (accessed 18th December 2021)

Catling, S. and Martin, F. (2011) Contesting powerful knowledge: The primary geography curriculum as an articulation between academic and children's (ethno-) geographies. *Curriculum Journal* 22(3), pp. 3 17–335.

Catling, S. and Pike, S. (2022) Becoming acquainted: Aspects of diversity in younger children's geographies In Hamond, L., Biddulph, M. Catling, L. and McKendrick, J. H. (eds.) *Children, Education and Geography: Rethinking Intersections.* Abingdon: Routledge

Chatterton, P. (2010) The student city: An ongoing story of neoliberalism, gentrification, and commodification. *Environment and Planning A* 42(3), pp. 509–514.

Christie, H. (2007) Higher education and spatial (im) mobility: Nontraditional students and living at home. *Environment and Planning A* 39, pp. 2445–2463.

Decolonising Geography (2021) *About.* https://decolonisegeography.com/about (accessed 18th December 2021)

Fielding, M. and Rudduck, J. (2002) The transformation potential of student voice: Confronting the power issues. Contribution to the symposium '*Student Consultation, Community and Democratic Tradition*' Presented at the Annual Conference of the British Educational Research Association, University of Exeter, 12–14 September 2002.

Firth, R. (2013) What constitutes knowledge in geography. In Lambert, D. and Jones, M. (eds.) *Debates in Geography Education.* Abingdon: Routledge.

Firth, R. (2018) Recontextualising geography as a school subject. In Jones, M. and lambert, D. (eds.) *Debates in Geography Education*, 2nd edition. Abingdon: Routledg.

Geographical Association (n.d.) *Thinking geographically.* www.geography.org.uk/Thinking-geographically (accessed 31th October 2021).

Grant, B. (1997) Disciplining Students: The construction of student subjectivities. *British Journal of Sociology of Education* 18(1), pp. 101–114.

Haigh, M.J. and Kilmartin, M.P. (1999) 'Student perceptions of the development of personal transferable skills. *Journal of Geography in Higher Education* 23(2), pp. 195–206.

Hammond, L. (2020) *An investigation into children's geographies and their value to geography education in schools.* Thesis (PhD). London: University College London.

Hammond, L. and Healy, G. (2022) 'Student voice, democratic education and geography: Reflecting on the findings of a survey of undergraduate geography students'. In Hammond, L. Biddulph. M. Catling, S. McKendrick, J. H. (eds.) *Children, Education and Geography: Rethinking Intersections.* Abingdon: Routledge

Herbert, D.T. and Thomas, C. J. (1998) School performance, league tables and social geography. *Applied Geography* 18(3), pp. 199–223.

Holdsworth, C. (2006) 'Don't you think you're missing out, living at home?' Student experiences and residential transitions. *The Sociological Review* 54, pp. 495–519.

Holdsworth C. (2009a). 'Between two worlds: local students in higher education and 'Scouse'/student identities'. *Population, Space and Place* 15(3), pp. 225–237

Holdsworth C. (2009b). 'Going away to uni': Mobility, modernity, and independence of English higher education students'. *Environment and Planning A*, 41(8), pp. 1849–1864.

Holton, M. and Riley, M. (2013) Student geographies: Exploring the diverse geographies of students and higher education. *Geography Compass* 7(1), pp. 61–74.

Hopkins, P. (2010) *Young People, Place and Identity*. London: Routledge.

Hopkins, P. (2014) Managing strangerhood: Young Sikh men's strategies. *Environment and Planning A* 46, pp. 1572–1585.

Hopkins, P. (2020) Social geography III: Committing to social justice. *Progress in Human Geography*, 45(2), pp382–293.

Hopkins, P, (2021) *Scotland's Islamophobia: Report of the Inquiry into Islamophobia by the Cross-Party Group on Tackling Islamophobia*. Newcastle upon Tyne: Newcastle University.

Hopkins, P., Botterill, K., Sanghera, G. and Arshad, R. (2015) *Faith, Ethnicity, Place: Young People's Everyday Geopolitics in Scotland*. Newcastle University: Newcastle upon Tyne.

Hopkins, P., Botterill, K., Sanghera, G. and Arshad, R. (2017) Encountering misrecognition: being mistaken for being Muslim. *Annals of the American Association of Geographers* 107(4), pp. 934–948

Hopkins, P., Botterill, K. and Sanghera, G. (2018) Towards inclusive geographies? Young people, race, religion and migration. *Geography* 103(2), pp. 89–95.

Hopkins, P., Sinclair, C. and Student Research Committee. (2017) Research, relevance and respect: Co-creating a guide about involving young people in social research. *Research for All* 1(1), pp. 121–127.

Jackson, P. (2006) Thinking geographically. *Geography*, 92(3), pp. 199–204.

Johnson, F., Fryer-Smith, E., Phillips, C., Skowron, L., Sweet, O. and Sweetman, R. (2009) *Raising Young Peoples Higher Education Aspirations: Teachers' Attitudes*. Department for Innovation, Universities and Skills, Report No 09 01. Available at: https://core.ac.uk/download/pdf/4158341.pdf (accessed 1th November 2021).

Jones, M. and Fitzgerald, B. (2007) Landscape of language: Geography across the curriculum. *Teaching Geography*, 32(1), pp. 22–28.

Korkiamäki, R. and Gilligan, R. (2020) Responding to misrecognition – A study with unaccompanied asylum-seeking minors. *Children and Youth Services Review*, 19, pp. 1–9.

Kraftl, P. (2013) *Geographies of Alternative Education: Diverse Learning Spaces for Children and Young People*. Bristol: Policy Press.

Lambert, D. (2011) Reviewing the case for geography and the'Knowledge Turn' in the English National Curriculum. *Curriculum Journal* 22(2), pp. 243–264.

Lander, V. (2015) 'Racism, its part of my everyday life' Black and minority ethnic pupils experiences in a predominantly white school. In Alexander, C, Weekes-Bernard, D. and Arday, J (eds.) *The Runnymede School Report: Race, Education and Inequality in Contemporary Britain*. London: Runnymede Trust, pp. 32–35.

Maguire, S., Evans, S.E. and Dyas, L. (2001) Approaches to learning: A study of first-year geography undergraduates. *Journal of Geography in Higher Education* 25(1), pp. 95–107.

Marriott, A. (2007) The transition from A level to degree geography. *Teaching Geography* 32(1), pp. 49–50.

McKendrick, J.H. (2022) Children's geographies and schools: Beyond the mandated curriculum. In Hammond, L., Biddulph, M., Catling, S. and McKendrick, J. H. (eds.) *Children, Education and Geography: Rethinking Intersections*. Abingdon: Routledge

Ofsted (2021) *Research review series: Geography*. www.gov.uk/government/publications/research-review-series-geography/research-review-series-geography (accessed 3th November 2021)

Pötschulat M., Moran M. and Jones P. (2021) The student experience' and the remaking of contemporary studenthood: A critical intervention. *The Sociological Review* 69(1), pp. 3–20.

Public Accounts Committee (2018) *The Higher Education Market: Forty-Fifth Report of Session 2017–19*. London: The Stationary Office.

Reay, D., Davies, J., David, M. and Ball, S.J. (2001) Choices of degree or degrees of choice? Class, 'race' and the higher education choice process. *Sociology* 35, pp. 855–874.

Roberts, M. (2013a) *Powerful knowledge: A critique.* www.youtube.com/watch?v=DyGwbPmim7o (accessed 13th July 2021).

Roberts, M. (2013b) *Geography through Enquiry: Approaches to Teaching and Learning in the Secondary School.* Sheffield: Geographical Association.

Smith, H.J, Robertson, L.H, Auger, N. and Wysocki, L. (2020) Translanguaging as a political act with Roma: Carving a path between pluralism and collectivism for transformation. *Journal for Critical Education Policy Studies* 18(1), pp. 98–135.

Tate, S. and Hopkins, P. (2020) *Studying Geography at the University: How to Succeed in the First Yar of Your New Degree*. London: Routledge.

Tate, S. and Swords, J. (2013) 'Please mind the gap: Students' perspectives of the transition in academic skills between A-level and degree-level geography'. *Journal of Geography in Higher Education* 37(2), pp. 230–240.

Taylor, C. (1994) The politics of recognition. In Taylor, C., Appiah, K.A., Habermas, J., Rockerfeller, S.C., Wlazer, M. and Wolf, S. (eds.) *Multiculturalism,* Princeton, NJ: Princeton University Press, pp. 25–74.

Taylor, C. (2009) Towards a geography of education. *Oxford Review of Education* 35(5), pp. 651–669.

Taylor, L. (2008) Key Concepts and medium-term planning. *Teaching Geography* 33(2), pp. 50–54.

Thomson, P. and Gunther, H. (2006) From 'consulting pupils' to 'pupils as researchers': A situated case narrative. *British Educational Research Journal* 32(2), pp. 87–103.

Tierney, W.G. and Venegas, K.M. (2006) Fictive kin and social capital: The role of peer groups in applying and paying for college. *American Behavioural Scientist* 49(12), pp. 1687–1702.

United Nations (1989) *Convention on the Rights of the Child, Treaty no.27531. United Nations Treaty Series,* 1577, pp. 31–178. https://treaties.un.org/doc/Treaties/1990/09/19900902%2003-14%20AM/Ch_IV_11p.pdf [Accessed 1.11.2021]

Waters, J. and Brooks, R. (2022) 'Geographies of education at macro-, meso and micro-scales: young people and international student mobility'. In Hammond, L., Biddulph, M. Catling, S. and McKendrick, J. H. (eds.) *Children, Education and Geography: Rethinking Intersections*. Abingdon: Routledge

Xiaowei, X., Qingna, J., Injeong, J., Yushan, D. and Mijung, K. (2019) The potential contribution of geography curriculum to scientific literacy. *Journal of Geography* 118(5), 185–196. DOI: 10.1080/00221341.2019.1611906

6

BECOMING ACQUAINTED

Aspects of diversity in children's geographies

Simon Catling and Susan Pike

Introduction

Children's geographies are diverse, multi-layered, and evolving. Even in the same family, no two children's geographies are identical; family position, ages, and relationships affect changes of access to and engagement with places. Their personal geographies change as they grow. Generally, close and restricted in their early years, during their pre-teens children's geographies evolve. Family and societal perspectives and circumstances, as well as engagement with physical activities and digital resources develop their geographies. They develop experience and knowledge about and engagement with the local and extended world, which may result in greater freedom or, conversely, control and restraint in their lives. What is possible for them is affected by opportunities and constraints, including their own perceptions and ideas about people and places. These factors impact on their knowledge, interests, values, and capabilities, affecting how they use their environments and understand the world beyond. With so many dynamics around children's geographies, education, and school geography, we focus on children up to 12 years old.[1]

Many dimensions of children's geographies have been researched, including schools, homes, neighbourhoods, and communities (Holloway and Valentine, 2000; Fog Olwig and Gulov, 2003; Evans and Horton, 2016; Freeman and Tranter, 2011). We explore some pertinent findings about children's place experiences, spatiality, geographical knowledge, and environmental awareness. We also consider their opinions on places and associated emotions, including ways that local experience contributes to identities and community participation. Immediate and local places are where children become acquainted with their personal worlds and senses of being and belonging.

Children's immediate neighbourhood is a third, lived space where their cultural identities are forged, mapped, and performed (Matthews et al., 2000). *Thirdspace*

DOI: 10.4324/9781003248538-9

is conceptualised as where 'everything comes together': subjectivity and objectivity, the repetitive and differential, structure and agency, and 'everyday life and unending history' (Soja, 1996: p. 57). For children, thirdspace involves their homes, schools, neighbourhood, and related places, real and virtual, considered, reflected upon, and augmented by their thoughtful geographies and through their learning in school curriculum geography. We consider the complexity of diversity in children's geographies. To unpick this complexity, we explore the variety and range of children's geographies through:

- children's geographies of places – the impact and importance of the local;
- children's thoughtful geographies of hope and challenge – connection and pleasure, concern and justice;
- children's learning and geographies – encountering geographies in school.

Children's geographies of places

There is much that can be considered about children's geographies in and of places. We explore the geographies of their immediate lives. We begin with children's home space, as their family's base-place, before considering aspects of their neighbourhood, or local, experience, within which one differentiated and controlled context is their school. Home is their first geographical physical and social environment and the neighbourhood (with school) their next. To do children's engagement in spaces a modicum of justice, we have limited ourselves to these places and contexts in which their geographical experience is initiated, explored, and fostered.

Home as child place and space

For the vast majority of children, home is where first they become acquainted with place, naming it 'home'. Whether stand-alone or apartments, homes vary in style and size, as well as facilities such as outdoor space. Whilst most homes are houses or flats, still many children, mostly not through parental choice, live in spaces not designed as homes, such as trailers, tents, and hotel rooms or are accommodated by other family members or friends. For some children home is less tangible, even unrecognisable, due to residential placements for themselves and/or those who care for them.

Before they become mobile, babies and children notice movement and features, initiating their sense of an environment and of it as a place and space (Orrmalm, 2021). Crawling and toddling around their home enable very young children's learning of features and locations (Newcombe and Huttenlocher, 2000). As Green (2018) notes, play in familiar places, under beds or in constructed indoor dens, ensures imaginative meanings. This knowledge of the names, locations, shapes, and functions of home spaces and places develops early spatial awareness and environmental geographies interplaying with language skills and initial vocabulary. The variety of ways children live, do jobs, and play in (and outside) their homes enables them to create personalised geographies and apply these in other places when visiting friends or pre-school centres. This understanding of the nature of home spaces, what they offer, and how children use them is essential in appreciating children's home geographies (Maitland et al., 2019).

A key site in homes for children is their bed space or room. Even where shared, perhaps crowded, it may be their significant thirdspace (Adcock, 2016). But these children's geographies have been little studied (Bacon, 2018). An adult-moulded arena, children often modify the look of their space, however small, by using items they collect to brand it as theirs, noting when it is respected or violated. For some children, bedrooms, as some homes, are places of escape. Yet bedrooms can be unsafe spaces, sites of abuse (Hörschelmann, 2017), and trauma with long-lasting effects, where children may equally gather the resilience to survive. Homes, therefore, have a diverse range of meanings among any group of children, some keeping secrets while others happily share their home experiences and feelings.

Home is a particularly meaningful notion for children who have moved home. For children home may include the neighbourhood where the family lived before moving. Early memories may be warmly imprinted. This is often so for families who have changed their residence locally or nationally for housing, employment, or other reasons, with children feeling they have left 'home'. For migrant and refugee children a new home in a new country is likely to be very different, perhaps initially an asylum centre or temporary hostel until their status is settled in their new country. Children have concerns about asylum accommodation, including its overcrowded environment and exclusion from social, education, or play and leisure activities (Arnold, 2015). Archambault (2012) noted that asylum children hoped for 'good-sized' space when they were rehoused but soon realised they had negligible choice and must live, to start with, where they were allocated. This affected whether they felt 'at home', despite families acquiring items to help create a sense of home. It can take some time for children and parents to begin to feel another place becomes home. Children who move regularly between the different homes and neighbourhoods of their separate parents can take time to build this sense of belonging. It will be likewise for a young child going into a children's home or with an adopting family, where a new 'home' place for some may be a 'not home' for others (Callaghan et al., 2016).

Children's sense of home depends on personal circumstances. Their family and residential circumstances affect how they identify with their home and its surroundings. While many children may have much in common, their experiences in homes and their notions of home are diverse. We know that it cannot be assumed that children even in the same group, at any age, have similar experiences of home. These affect their environmental and spatial home geographies, which can in turn impact on their experiences in and appreciation of their neighbourhoods.

Neighbourhood as a child's locale

Most children live much of their lives in a particular locality, their neighbourhood. Early in life, children's awareness of local neighbourhoods is parentally controlled, but later local travels develop their physical skills, spatial awareness, and knowledge in and of local places. Concerns have long been raised about children's declining use of the local outdoors (Louv, 2005), but their experiences worldwide are mixed living in places they and their families love to be in or are concerned about when they go outside. Many children have always lived in the physically risky landscapes of earthquakes, tropical storms, and flooding yet spend much time in their neighbourhood outdoors (Chatterjee, 2017, 2018).

Generally, children under six spend their time in outside spaces nearby home with parents and other adults, such as their grandparents or employed caregivers, who select sites with which they are comfortable. In some cases and safe contexts, young children may be able to explore the immediate area around home with siblings and occasionally unaccompanied. However, getting to know the wider proximate area and the neighbourhood through personal movement tends to begin from 7–8-years-old but may not occur until later, depending on carer's perspectives of the locale. Children's senses of belonging and identities are created through exploring the affordances of local places, often with others (Evans and Horton, 2016; Freeman and Tranter, 2011). Yet, as Loebach and Gilliland (2019) noted conversely, where children interact less with their locality, they display a negligible sense of neighbourhood belonging. This indicates the significance of active, stimulating local engagement with affordances to create meaning (Pike, 2011). The evidence suggests that for children of all ages engaging with and valuing a place or specific aspects of it provides them with a sense of attachment which enhances their feeling of security in their place, in their use of their home range, and in building their local knowledge and identification with it.

Though neighbourhoods are adult-centred, children appropriate them as cultural thirdspaces within which they are participants negotiating their influence and adapting places for their own interests. This may occur in places designed for children (such as playgrounds and trampolining or soft play spaces) or in larger environs children use (for example, parks or green areas). Such uses and appropriation help create deeper complexities in neighbourhoods, at times within but also beyond adults' gaze (Cobb et al., 2005). This can involve subverting purposed spaces, perhaps a site in a wild or wooded area for den creation. Neighbourhoods become arenas of meaning as social meeting sites by like-minded peers, such as migrant children (De Visscher, 2014). Children's free-drawn maps invariably include such valued places. Owens et al. (2020) noted young children demonstrating *core* and *cultural* knowledge, or 'designative' and 'appraisive' elements (Salameh and Çubukçu, 2018), on their maps. *Core* knowledge names places and common features. In their maps these children delineated small local areas, in which they depicted few natural features. *Cultural* knowledge is affective and evaluative, reflecting personal meaning, such as places identified with or where events and discoveries had been made. These varied child to child. Prakoso (2018) found 9–12-year-olds' favourite sites in a relatively non-child-friendly city were shops, play areas, trees, parks, benches, and friends' homes. Such 'lived-existential' thirdspaces provided opportunities for children to enact their neighbourhood agency with personal meaning and value through activities, purchases, and meeting others (Freeman and Tranter, 2011; Pike, 2011, 2020).

Through such personal geographies, children develop perspectives on their neighbourhoods, their likes, dislikes, and issues, which they can describe, explain, and sometimes act upon (Pike, 2016, 2021). While local places are not always seen positively (Hayball et al., 2017), and children may not always have a strong affinity with their neighbourhood, they know the local risks and dangers and use avoidance strategies to minimise these (Jamme et al., 2018). Wilson et al. (2019) found 10–12-year-olds identified different important features to adults' during their walks to school, which they perceived as enablers (such as parks with trees to explore and climb) and barriers (poor weather and missing pavements). Walking with friends and siblings was a safety enabler supporting journeys along busy streets or through littered sites. Children are impacted by adult social attitudes, as in being seen as 'out of place' in liminal urban spaces (Valentine, 2004). Often, distant structural global changes can affect children's personal local geographies, similarly negatively (Aitken, 2001), as Katz (2004) in Sudan and Abebe (2007) in Ethiopia found when younger children's local work patterns increased and changed, resulting from international trade inequalities and development projects.

Evidently, neighbourhood matters in younger children's lives, development, and health (Minh et al., 2017). Places make a difference with children but not necessarily quite as assumed. Inequality and diversity are influences in neighbourhoods, even when 'hidden', that provide constraints on and freedoms for children. Relatively affluent and poor neighbourhoods often contain influential variations, related to income, available opportunities, and social perceptions (Hooper et al., 2007). These patterns are complex and influenced by many factors which in turn affect local norms of accepted activity. For instance, some children in wealthier areas may be discouraged from spending time outside home, while those from low-income families may make fuller use of local leisure facilities and activities because of community provision and support (Pike, 2011). During the 2020–2021 COVID-19 restrictions, patterns shifted, with only 19% of children from the highest but 36% from the lowest income groups spending less time outside (ESRI, 2021). Government restrictions temporarily closed many leisure and commercial facilities and barred or discouraged families and friends mixing, resulting inevitably in less outdoor activity, even though children were perceived as less at risk than adults from the disease.

The use of the neighbourhood is significant, as it makes up 'liminal spaces where both the transformative and mundane occur' for children (Pimlott-Wilson and Hall, 2017: p. 259). Neighbourhoods have lifelong effects, enabling and inhibiting life chances (Chetty et al., 2018). Early developing experiential geographies are significant influences during children's young lives; it is important they and those who care for them at home and in education and the wider community understand and appreciate this.

School as 'child place/space'

In early years primary school is a significant lived place; most children get to know their school's physical and social geography. Globally, children have very varied experience of pre-schools and schools, from the state-of-the-art, spacious, and well-resourced to the dilapidated, crowded, and poorly financed, from varied grounds to a lack of play space and from full- to half-day attendance with or without well-qualified early years and primary teachers. While some children have access to child minders, play groups, nurseries, and even live-in carers, across the world, as many as 20% of children, particularly girls, have little or no experience of school (UNICEF, 2021). School is not a given for every child.

For very many children, school, as a place, matters; school is 'my school', to which they have a sense of belonging (Rieh, 2020). This is also true of 'my child minder's home' and 'my after-school club' whether these are formal paid-for places or arrangements through family or friends. Many children have highly positive experiences of education, enjoying social and learning engagement with their peers and teachers, effectively using the affordances of the school. Yet for others, or at other times, school may be a place with negative associations, through bullying, lack of voice, or being sanctioned. For some children, their experiences in one school can include all these, although more research is needed into early years and primary school places.

Classrooms are spaces for curriculum learning, and they are sites in which children learn about the impact of space and layouts, social interactions, and power relations. The youngest children in school recognise how classrooms are controlled and what leeway they have (Dixon, 2004). Barnikis (2015) noted that children know about the division and control of space in school buildings. As adult-controlled spaces, children learn about access to and the uses of spaces, from the teacher's desk and the resource cupboard to prearranged classroom layouts and rules for using reading and craft areas, though children with mobility challenges (Stephens et al., 2017) can encounter access difficulties. While children learn that adults regulate pre-school and school spaces through monitoring (Devine, 2003; Millei, 2018), they also identify thirdspaces, such as the toilets, where they might take a 'break', sensing teachers see these as unimportant sites, offering opportunities for children's agency, for instance, by taking 'more' time over ablutions, chatting to whoever happens to be there at the same time, or walking more slowly to and from the toilet. By observing adults monitoring and surveilling classrooms, playgrounds, and lunch spaces, children realise how they might use different sites to gain some control over these lived spaces for their own purposes (Kellock and Sexton, 2018). And nurture and sensory rooms in schools often include elements of design by children but may not be used by all children.

School play areas offer possibilities and limitations for younger children. In them children may feel central or peripheral, based on age, race, religion, disability, gender, or physical skills (Ndhlovu and Varea, 2018). Feelings of being socially excluded or bullied may lead some children to seek solace in 'their own space' (Brown,

2017), with others enjoying being in small groups separate from 'the crowd', while bigger groups, often boys, take over large spaces for ball games. Effectively, these are child-made decisions, with children seemingly accepting expectations and norms for different activities allowing or constraining place use and related to physically exertive or calm activities (Ndhlovu and Varea, 2018).

Children often have favourite sites in school grounds, associated with feelings of being and/or belonging in a school. They have diverse connections with play spaces. Some hold a wider sense of the 'whole' space, wandering the site (Brown, 2017). Favourite sites enable children to feel free, safe, or hidden, to play and relax, and to expropriate features and sites for their discrete uses. (Un)conventional uses of places are part of children's play mix, from rule-based games to creating places for imaginary friends. Equally, children are aware of the limitations and benefits of favoured areas and have suggestions about improvements (Christidou et al., 2013), such as access to wild areas, which teacher control is perceived to limit (Groves and McNish, 2011). Though playgrounds may offer some variety through their spaces and features, younger children find ways to adapt or repurpose these liminal spaces, for their own engagement, ascribing them personal meanings.

Children's thoughtful geographies of hope and challenge

Children's geographies include their thoughts, opinions, and ideas about places. These are their complex and shifting 'thoughtful geographies', the delights, despairs, hopes, and fears of children growing up in the Anthropocene (Malone, 2018). They are dynamic for individuals, shaped by the shifting nature of nat-ural and built neighbourhoods, as well as by the changing spatial relations and power asymmetries manifested in places (Aitken, 2001). These impact on children's lives and futures (Catling, 2003) and will include during their childhood 'a range of newly emerging phenomenon not yet understood or anticipated' (Comber, 2012: p. 4–5). Children's thoughtful geographies, emergent through their lived experiences, agency, and thoughts about the future, are presented as continua in Figure 6.1. The figure simplifies the complex ranges of children's current positivity about delights and concerns that may lead to despair(s), as much as their hopes and fears for the future. We use this diagram to illustrate children's shifting thoughts.

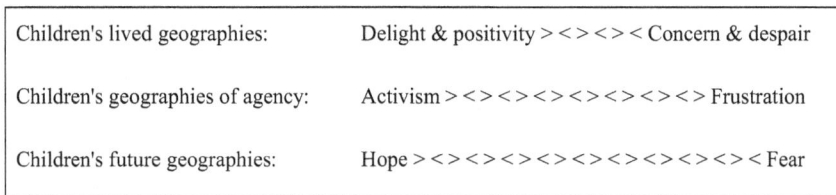

Children's lived geographies:	Delight & positivity > < > < > < Concern & despair
Children's geographies of agency:	Activism > < > < > < > < > < > < > Frustration
Children's future geographies:	Hope > < > < > < > < > < > < > < > < > < Fear

FIGURE 6.1 A continuum of children's thoughtful geographies

Note: > < > < > < denotes that children's thoughtful geographies move back and forward, dependent on their contexts, emotions, perspectives, and reactions.

For example, across a single day, children's thoughts on their lives and futures can shift and change according to what they see, hear, and think about at home and in school, through news and social media, conversations, and what they are taught. Children's thoughts, opinions, and ideas are significant since they affect so many aspects of their lives, including their education. Children's, and adults', actions and frustrations can move children's lived experiences back and forth across these continua to increase delight and hope, or fear and despair, such as about climate change, biodiversity, and care for and in local communities.

Children delight in harnessing their agency to create opportunities to play, move around, and interact, talking positively of their experiences in their neighbourhood while being frustrated at issues like litter. Children value the risks they take in their locality and see these as part of childhood (Yates and Oates, 2019). However, there are dangers children may be deeply concerned at, such as anti-social activities that wake them at night or graffiti that generate feelings of being unsafe (Pike, 2011). Even the youngest children can articulate their views about places (Yates and Oates, 2019). They may express frustration where they cannot act for change, when there is little compassion or justice for themselves and/or others. For some children features they have no control over lead to feelings of despair, for example, in the travelling community when children living on halting sites may lack basic rights to homes, play, water, and education due to weak housing provision (OOC, 2021); this is not surprising since their basic human rights remain not met. This is equally true for many other children across the variety of places, where some may lack a home and any opportunities for play and education, as is the case for millions of girls in many countries and can be for migrants and asylum seekers.

While they may not immediately have the language to express it, children seem to recognise the impact of others, such as decision-makers, on their geographies (Schlemper et al., 2018). Yet their concerns and frustrations can be mobilised into delight, as they show positive desires and intentions, especially with adult support from their educators (Hart, 1997; Shier, 2001). When contributing to the community through school-based projects, children can create positive changes in their lives. For example, children from Central Model (Primary) School, Dublin, lobbied local government successfully against the closure of their local swimming pool. On national television, they expressed pride in their efforts and encouraged others to follow their example (RTÉ, 2021), demonstrating the impact of child agency and activism with teacher support.

Children's thoughts about their futures are as varied as their ideas about their current lives, influenced in part by their family, friends, and school within their neighbourhood. As Williams (2021) notes, in examining place, poverty, and learning in Wales's Rhondda valleys, these directly affect the lived experiences in place of children, who have high expectations for their futures, clearly challenging local policy presumptions. Even where they are anxious, Hickman (2020, 22) notes this is an 'emotionally congruent healthy response' to what they learn. Overall, it appears children have advanced and evolving knowledge of current circumstances in and futures for their communities and localities. Thew et al. (2020) found young participants in climate strikes first articulated injustices relating to generational future risks but over time developed perspectives about injustices experienced by other groups in the present. Children's knowledge can influence adults' views on the Earth's future, such as about climate injustices, as they discuss what they have learnt at home (Lawson et al., 2019). Adults should support internal and external activism in building resilience in children and supporting their campaigns for change. Hope is a vital factor to be promoted in children's lives (Dolan, 2022), illustrating one of many possible shifts from concern to positivity in the continua shown in Figure 6.1. But more research is required, especially about how thoughtful geographies affect younger children's lived geographies.

Children's learning and geographies

Younger children's lives develop essential and meaningful connections through their local geographies (Barlow and Whitehouse, 2019). Opportunities to enhance and build on children's geographies through experiences and learning can and should occur at home, within communities, and at school (Catling and Martin, 2011).

A key dimension of primary schooling is learning how to act in 'public' places, to know what is expected, in terms of actions and behaviour. Schools are governed spaces, regulated so that children learn what the curriculum requires and how to use space and accept values that show these places matter to them and others. Such caring for their school, physically, socially, and ethically, values their geography of place (Fielding, 2000). While at times explicit, this is often an implicit or hidden curriculum, until transgressed, in daily activities and routines in classrooms, corridors, and playgrounds. Largely, children follow the rules but are capable of using and subverting a school's liminal spaces as personal sites, for instance, so as to avoid peers to read while others find suitable places to converse. Children learn and understand how their playgrounds work; even those children playing sport during recesses can provide accounts of wider playground space uses. These are inevitable dimensions of their school-based informal geographical learning.

Early years and primary curriculum geography and geographical experiences in schools and other educational settings provide children with opportunities to learn about people, natural and human features, and environmental processes at a range of scales. Where taught well, with engaging activities and appropriate challenge, it is a subject that is valued and liked by children (Pike, 2016; Dolan, 2020). Importantly, such learning develops knowledge and enables enlightenment. Primary geography has a significant role to play by drawing on children's geographies to value, involve, and take children deeper into and beyond their lived and experienced geographies (Catling and Martin, 2011). One key approach involves children studying their locality to extend and deepen their knowledge, become more aware of how their place has multiple geographical dimensions, and foster their sense of identity with and belonging to 'their place'. This extends children's understanding of their personal geographies. The deeper engagement with their neighbourhood provides for them a relevant way into 'placemaking' about their area (Derr et al.,

2018) and into opportunities to be involved in environmental change, exploring matters affecting them within and from beyond their places, such as the climate crisis (Dolan, 2022). In this way, formal geographical learning, by drawing on children's lived geographies, becomes a joint endeavour of children and teachers, who come to value children's agency as experience-based learners of curriculum geography (Pike, 2011; Hammond, 2022; Biddulph et al., 2022). Children can enable their learning to be reflected in their communities through active contributions.

Many argue that children's engagement and activism enable social change and transformation in educational and community settings, for instance, in improving school sustainability and city-friendliness for children and adults (Dolan, 2020; Kavanagh et al., 2021; Malone, 2018). Such thinking is linked to children's rights to participate in their local communities on matters which affect them (Shier, 2001), as increasingly research evidences (McMahon et al., 2018). Community- and school-based research projects have noted that children, supported by adults, are capable of advanced thinking about possible and desired futures (Hicks, 2014), for example, in identifying practical ways to green their neighbourhood. Children's enthusiasm for such projects fosters a sense of community, accomplishment, and empowerment (Torres-Harding et al., 2018). Through supported geographical enquiry using fieldwork techniques and digital technologies, children in one project understood the structural dimensions of the frustrations faced in the community and explored ways forward (Schlemper, et al., 2018), later presenting their findings and recommendations to community stakeholders. Such projects indicate ways in which younger children can interlink and enhance their opportunities for geographical learning with the experiential dynamics and priorities in their own geographies (Hammond, 2022). They indicate the importance for primary teachers of recognising and engaging with children's and local geographies, to extend their own 'funds of knowledge' for teaching by engaging with their children's geographies (Catling and Willy, 2018).

Engaging children's geographies within school leads to a range of authentic and motivating learning (Scoffham, 2016). We argue that children learn their geographies vicariously. Formal curriculum learning is the main intended activity for which schools exist, but it is not the only learning of 'geography' which takes place in and outside primary schools, as we have indicated. It is the place of educators to make professional judgements about the opportunities in and limits of school learning for children. It can be that in their geography curriculum, children identify and discuss controversial matters. When children raise personal or societal issues that they argue need addressing, should the learning that emerges lead to activism? It can do, as the climate crisis has shown. In recent years, many children have acted in relation to their geographies, through feelings of despair for the future, by striking, local actions, and contacting politicians about their concerns with the climate crisis and the inadequate responses from decision-makers in government and the corporate world (Dolan, 2022). They demonstrate their desire for a better future.

Opportunities for children's geographical learning occur in non-school educational contexts. One example is through the 'forest school' approach, whether school-organised or outside school (Kraftl, 2013; Knight, 2013; Williams-Siegfredsen, 2017); another is beach-, river-, and pond-based 'water school' (Horvarth, 2016). Children enjoy opportunities to explore and investigate the natural world, using play and personal investigation as ways to learn about the nature of environments. Woodland and water sites encourage early and primary years' children to ask questions about and to consider and respect places and their ecosystems. These approaches to outdoor learning (Waller et al., 2017; Waite, 2017) motivate many children and develop their knowledge and appreciation of the natural world and their relationship with it (Ridgers et al., 2012; Martens et al., 2020). They can draw on and extend children's experiences in, and awareness and understanding of environments, as well as foster, for instance, their self-esteem, capacity to risk-take, team-working, and vocabulary and communication skills (Austin et al., 2016; Knight, 2017). These, as in other contexts beyond schooling, such as the guides and woodcraft movements, enable children to extend their personal geographies of places and geographical learning in educational circumstances they may perceive as less formal and supervised.

In school, the primary geography curriculum provides opportunities for formative learning experiences, such as fieldwork and enquiry approaches, to enable children to act with delight for hope in their own schools and communities. At times this can reveal a far greater wish to be involved than the curriculum or resources they are presented with. For instance, in 2021, the UK government (HMG, 2021) provided various educational resources and suggested children could make posters using prepared slogans to 'celebrate' its 'efforts' on climate change (Figure 6.2). This simplistic approach illustrates how trivial resources and tasks produced by adults underestimated children's ability to think critically, imaginatively, and actively about the geographies affecting their communities and its wider interests and, indeed, affecting the world. Rather more informative and effective is using a range of strategies, including enquiry-based fieldwork, mapping, mysteries, and debating scenarios for the future, to investigate the possible impacts of climate change, solutions to plan for, and actions to take (Pike, 2016; Willy, 2019; Dolan, 2022), locally and more widely. Research into effective sustainability education shows geography contributes strongly. For example, Monroe et al. (2019) noted that sustainability education was most effective where it focused on personally relevant and meaningful information and used active and engaging teaching methods, all features of high-quality geography lessons (Ofsted, 2021). Where children successfully and meaningfully participate in school learning that combines learning with activism, research shows they gain in learning, including geographical learning (Pike, 2021).

Conclusion

Children's geographies are personal and varied. While children living in the same neighbourhood and attending the same educational settings share common ground, their geographies nonetheless are individual and different. The

FIGURE 6.2 Suggested slogans for children to use to support COP26

Source: HMG (2021)

A LOT OF IT
IS DOWN TO
EVERYDAY PEOPLE
MAKING CHANGES

#TOGETHERFOROURPLANET #COP26

AHEAD OF COP26,
JOIN US TO
CELEBRATE THEIR
INSPIRATIONAL
EFFORTS

#TOGETHERFOROURPLANET #COP26

FIGURE 6.2   *(Continued)*

lives and place contexts of children vary locally, nationally, and globally. Children resident on the same street, even in the same block of flats or collection of temporary homes, have differing experiences of 'their' place, as they have within their homes and schools. This diversity is not based in equality of access, fairness, and justice but is created by local, national, and international poverty and wealth, the nature and quality of their housing and neighbourhoods, the schools they attend, and their educational provision, political decision-making, and family and societal interests, perceptions, expectations, and priorities. Home, neighbourhood, school, and other educational places affect children, contributing to making their geographies, which intrinsically they create for themselves. Children's geographies are individual-, social-, place- and environment-based, offering them affordances and constraints, opportunities for familiarity and adventure, supporting senses of local identity and belonging, and enabling them to care and, quite possibly, contribute to have an impact. Yet children are normally marginalised, even in schools, surveilled and regulated overtly and implicitly. Children from their earliest years develop knowledge of the world through cyclical daily actions, observations, discussions, reflections, and understandings. They are agents in their own geographical learning at school and home, often with others, constantly building their knowledge bases and details about people, places, events, and the environment through experiences in and outside home and school.

Primary school geography has the power to support and enable children to make increasingly better and deeper sense of their understanding, particularly of the world beyond their immediate experience. When given time in schools, geography brings its particular way of looking at and thinking about the world to help children use its insights as they develop, offering varied stories and explanations about the world. It can provide opportunities for place learning, making, and enhancements. Children's worlds comprise multiple stories, enabling them to engage with various perspectives. Significantly, building children's sense of connection with their neighbourhood provides a key means to take into account their personal geographies, involving them in sharing these and discussing and debating what they know, how they perceive their locality, and their various views on ways to relate to and improve it. Primary school teachers need to recognise their children's lived geographies and integrate these into their geographical studies, not only as aspects of local geography but also in investigating human, physical, and environmental geography topics and global matters and concerns and places in the wider world. It means appreciating that younger children can – indeed, must – have agency in their own geographical learning.

Note

1   In line with the term 'children's geographies', as described in Chapter 1, applied to 0–11/12-year-olds.

References

Abebe, T. (2007) Changing livelihoods, changing childhoods: Patterns of children's work in rural southern Ethiopia. *Children's Geographies*, 5(1–2), pp. 77–93.

Adcock, J. (2016) The bedroom: A missing space within geographies of children and young people. In K. Nairn and P. Kraftl (eds.) *Geographies of Children and Young People 3: Space, Place, and Environment*. (pp. 401–420) Singapore: Springer Nature.

Aitken, S. (2001) *The Geographies of Young People: The Morally Contested Spaces of Identity*. Abingdon: Routledge.

Archambault, J. (2012) 'It can be good to be there too': Home and continuity in refugee children's narratives of settlement. *Children's Geographies*, 10(1), pp. 35–48.

Arnold, S. (2015) *State Sanctioned Child Poverty and Exclusion: The Case of Children in State Accommodation for Asylum Seekers*. Dublin: Irish Refugee Council.

Austin, C., Knowles, Z., Richards, K., McCree, M., Sayers, J. and Ridgers, N. (2016) Play and Learning Outdoors: Engaging with the Natural World Using Forest School in the UK. In K. Nairn and P. Kraftl (eds.) *Geographies of Children and Young People 3: Space, Place, and Environment*. (pp. 115–136) Singapore: Springer Nature.

Bacon, K. (2018) Children's use and control of bedroom space. In S. Punch and R. Vanderbeck (eds.) *Geographies of Children and Young People 5: Families, Intergenerationality, and Peer Group Pressure*. (pp. 85–105) Singapore: Springer Nature.

Barlow, A. and Whitehouse, S. (2019) *Mastering Primary Geography*. London: Bloomsbury.

Barnikis, T. (2015) Children's perceptions of their experiences in early learning environments: An exploration of power and hierarchy. *Global Studies of Childhood*, 5(3), p. 2910304.

Biddulph, M. Hopkins, P. Tate, S. (2022) 'Connecting children's and young people's geographies and geography education: Why this matters to and for children, education, and society. In L. Hammond, M. Biddulph, S. Catling and J.H. McKendrick (eds.) *Children, Education and Geography: Rethinking Intersections*. (pp. 69–82) Abingdon: Routledge.

Brown, C. (2017) 'Favourite places in school' for lower-set ability pupils: School groupings practices and children's spatial orientations. *Children's Geographies*, 15(4), pp. 399–412.

Callaghan, J., Fellin, L. and Alexander, J. (2016) Mental health of looked-after children: embodiment and use of space. In B. Evans and J. Horton (eds.) *Geographies of Children and Young People 9: Play and Recreation, Health and Wellbeing* (pp. 561–580) Singapore: Springer Nature.

Catling, S. (2003) Curriculum contested: Primary geography and social justice. *Geography*, 88(3), pp. 164–210.

Catling, S. and Martin, F. (2011) Contesting powerful knowledge: The primary geography curriculum as an articulation between academic and children's (ethno-) geographies. *The Curriculum Journal*, 22(3), pp. 317–335.

Catling, S. and Willy, T. (2018) *Understanding and Teaching Primary Geography*. London: Sage.

Chatterjee, S. (2017) *Access to Play for Children in Situations of Crisis: A synthesis of research in six countries*. Swindon: International Play Association. http://ipaworld.org/wp-content/uploads/2017/07/ipa-a4-access-to-play-in-situations-of-crisis-toolkit-lr.pdf

Chatterjee, S. (2018) Children's coping, adaption and resilience through play in situations of crisis. *Children, Youth and Environments*, 28(2), pp. 119–145.

Chetty, R., Friedman, J., Hendren, N., Jones, M. and Porter, S. (2018) *The Opportunity Atlas: Mapping the Childhood Roots of Social Mobility*. https://opportunityinsights.rg/wp-content/uploads/2018/10/atlas_paper.pdf.

Christidou, V., Tsevreni, I., Epitropou, M. and Kittas, C. (2013) Exploring primary children's views and experiences of the school ground: The case of a Greek school. *International Journal of Environmental & Science Education*, 8(1), pp. 59–83.

Cobb, C., Danby, S. and Farrell, A. (2005) Governance of children's everyday spaces. *Australian Journal of Early Childhood*, 30(1), 1pp. 4–20.

Comber, B. (2012) *Literacy for a Sustainable World*. South Melbourne, Australia: Oxford University Press.

De Visscher, S. (2014) Mapping children's presence in the neighbourhood. In: G. Biesta, M. Bouverne-De Brie and D. Wildemeersch (eds.) *Civic Learning, Democratic Citizenship and the Public Sphere*. (pp. 73–89) Dordrecht: Springer.

Derr, V., Chawla, L. and Mintzer, M. (2018) *Placemaking with Children and Youth: Participatory Practices for Planning Sustainable Communities*. New York: New Village Press.

Devine, D. (2003) *Children, Power and Schooling: How Childhood is Structured in the Primary School*. Stoke-on-Trent: Trentham Books.

Dixon, A. (2004) Space, schools and the younger child. *Forum*, 46(1), pp. 19–23.

Dolan, A. (2020) *Powerful Primary Geography: A Toolkit for 21st Century Learning*. Abingdon: Routledge.

Dolan, A. (ed.) (2022) *Teaching Climate Change of Primary Schools: An interdisciplinary approach*. Abingdon: Routledge.

ESRI (2021) *Growing Up in Ireland: Key findings from the special COVID-19 survey of Cohorts'98 and'08*. www.esri.ie/system/files/publications/BKMNEXT409.pdf

Evans, B. and Horton, J. (eds.) (2016) *Geographies of Children and Young People 9: Play, Recreation, Health and Wellbeing*. Singapore: Springer Nature.

Fielding, S. (2000) Walk on the left! Children's geographies and the primary school. In S. Holloway and G. Valentine (eds.) *Children's Geographies: Playing, living, learning*. (pp. 230–244). London: Routledge.

Fog Olwig, K. and Gulløv, E. (eds.) (2003) *Children's Places: Cross-cultural Perspectives*. London: Routledge.

Freeman, C. and Tranter, P. (2011) *Children and their Urban Environment: Changing Worlds*. London: Earthscan.

Green, C. (2018) Young children's spatial autonomy in their home environment and a forest school setting. *Journal of Pedagogy*, 9(1), pp. 65–83.

Groves, L. and McNish, H. (2011) *Natural Play: Making a Difference to Children's Learning and Well-being*. http://outdoorplayandlearning.org.uk/wp – content/uploads/2016/07/naturalplaystudyfull-1.pdf.

Hammond, L. (2022) Recognising and exploring children's geographies in school geography. *Children's Geographies*, 20(1), pp. 64–78.

Hart, R. (1997) *Children's Participation: The Theory and Practice of Involving Young Citizens in Community Development and Environmental Care*. London: Earthscan.

Hayball, F., McCrorie, P., Kirk, A., Gibson, A.-M. and Ellaway, A. (2017) Exploring children's perceptions of their local environment in relation to time spent outside. *Children & Society*, 32(1), pp. 4–26.

Hickman, C. (2020) We need to (find a way to) talk about . . . Eco-anxiety. *Journal of Social Work Practice*, 34(4), pp. 411–424.

Hicks, D. (2014) *Educating for Hope in Troubled Times*. London: Trentham Books/Institute of Education Press.

HMG [Her Majesty's Government] (2021) *UN Climate Change Conference UK 2021 Schools Pack: Resources*. https://together-for-our-planet.ukcop26.org/schools-pack-resources.

Holloway, S. and Valentine, G. (2000) *Children's Geographies: Playing, Living, Working*. Abingdon: Routledge.

Hooper, C.-A., Gorin, S., Cabral, C. and Dyson, C. (2007) Poverty and 'place': Does locality make a difference. *Poverty Journal*, 128, pp. 7–10.

Hörschelmann, K. (2017) Violent geographies of childhood and home: The child in the closet. In C. Harker and K. Hörschelmann (eds.) *Geographies of Children and Young People 11: Conflict, Violence and Peace.* (pp. 233–251) Singapore: Springer Nature.

Horvath, J. (2016) *Educating Children through Natural Water: How to Use Coastlines, Rivers and Lakes to Promote Learning and Development.* Abingdon: Routledge.

Jamme, H.-T., Bahl, D. and Banerjee, T. (2018) Between "broken windows" and the "eyes on the street": Walking to school in inner city San Diego. *Journal of Environmental Psychology*, 55, pp. 121–138.

Katz, C. (2004) *Growing Up Global: Economic Restructuring and Children's Everyday Lives.* Minneapolis, MN: University of Minneapolis Press.

Kavanagh, A., Waldron, F. and Mallon, B. (Eds.) (2021) *Teaching Social Justice and Sustainable Development across the Primary Curriculum.* Abingdon: Routledge.

Kellock, A. and Sexton, J. (2018) Whose space it is anyway? Learning about space to make space to learn. *Children's Geographies*, 16(2), pp. 115–127.

Knight, S. (Ed.) (2013) *International Perspectives on Forest School: Natural Spaces to Play and Learn.* London: Sage.

Knight, S. (2017) Forest School for the early Years in England. In T. Waller, E. Ärlemalm-Hagsér, E. Sandseter, L. Lee-Hammond, K. Lekies and S. Wyver (eds.) *The Sage Handbook of Outdoor Play and Learning* (pp. 97–110). London: Sage.

Kraftl. P. (2013) *Geographies of Alternative Education: Diverse Learning Spaces for Children and Young People.* Bristol: Policy Press.

Lawson, D., Stevenson, K., Carrier, S., Strannad, R. and Seekamp, E. (2019) Children can foster climate change concern among their parents. *Nature Climate Change*, 1(1), pp. 1–5.

Loebach, J. and Gilliland, J. (2019) Examining the social and built environment factors influencing children's independent use of their neighbourhoods and experiences of local settings as child-friendly. *Journal of Planning Education and Research*, doi/10.1177/0739456X19828444.

Louv, R (2005) *Last Child in the Woods: Saving our children from nature-deficit disorder.* Chapel Hill, NC: Algonquin Books.

Maitland, C., Foster, S., Stratton, G., Braham, R. and Rosenberg, M. (2019) Capturing the geography of children's active and sedentary behaviours at home: The HomeSPACE measurement tool. *Children's Geographies*, 17(3), pp. 291–308.

Malone, K. (2018) *Children in the Anthropocene: Rethinking Sustainability and Child Friendliness in Cities.* London: Palgrave Macmillan.

Martens, D. Friede, C. and Molitor, H. (2020) Natural experience areas: Rediscovering the potential of nature for children's development. In A. Cutter-Mackenzie-Knowles, K. Malone and E. Barratt Hacking (eds.) *Research Handbook on Childhoodnature: Assemblages of Childhood and Nature Research* (pp. 1469–1499). Cham: Springer Nature.

Matthews, H., Limb, M. and Taylor, M. (2000) The 'street as thirdsapce'. In S. Holloway and G. Valentine (eds.) *Children's Geographies: Playing, Living, Learning* (pp. 63–79). London: Routledge.

McMahon, G., Percy-Smith, B., Nigel, T., Bečević, Z., Liljeholm Hansson, S. and Forkby, T. (2018) *Young People's Participation: Learning from Action Research in Eight European Cities.* Huddersfield: University of Huddersfield. https://zenodo.org/record/1240227#.YRzm-9O2kRE

Millei, Z (2018) Generating educational and caring spaces for young children: Case of preschool bathroom. In S. Punch and R. Vanderbeck (eds.) *Geographies of Children and Young People 5: Families, Intergenerationality, and Peer Group Pressure* (pp. 287–306). Singapore: Springer Nature.

Minh, A., Muhajarine, N., Janus, M., Brownell, M. and Guhn M. (2017) A review of neighbourhood effects and early childhood development: How, where, and for whom, do neigbourhoods matter? *Health & Place*, 46, pp. 155–174.

Monroe, M. C., Plate, R. R., Oxarart, A., Bowers, A. and Chaves, W. A. (2019) Identifying effective climate change education strategies: A systematic review of the research. *Environmental Education Research*, 25(6), pp. 791–812.

Ndhlovu, S. and Varea, V. (2018) Primary school playgrounds as spaces of inclusion/exclusion in New South Wales, Australia. *Education 3–13*, 46(5), 494–505.

Newcombe, N. and Huttenlocher, J. (2000) *Making Space: The Development of Spatial Representation and Reasoning*. Cambridge, MA: MIT Press.

Ofsted [Office for Standards in Education] (2021) *Research Review Series: Geography*. www.gov.uk/government/publications/research-review-series-geography/research-review-series-geography

OOC [Office for the Ombudsman for Children (Ireland)] (2021) *No End in Site: An Investigation into the Living Conditions of Children on a Local Authority Halting Site*. Dublin: OOC.

Orrmalm, A. (2021) The flows of things – exploring babies' everyday space-making. *Children's Geographies*, 19(6), pp. 677–688. https://doi.org/10.1080/14733285.2020.1866748.

Owens P., Scoffham, S., Vujakovic, P. and Bass, A. (2020) Meaningful maps. *Primary Geography*, 102, pp. 15–17.

Pike, S. (2011) "If you went out it would stick": Irish children's learning in their local environment. *International Research in Geographical and Environmental Education*, 20(2), pp. 139–159.

Pike, S. (2016) *Learning Primary Geography: Ideas and Inspirations from Classrooms*. Abingdon: Routledge.

Pike, S. (2020) GIS for young people's participatory geography. In N. Walshe and G. Healey (eds.) *Geography Education in the Digital World: Linking Theory and Practice* (pp. 117–128). Abingdon: Routledge.

Pike, S. (2021) Geography education for social and environmental justice education. In. A.M. Kavanagh, F. Waldron and B. Mallon (eds.) *Teaching Social Justice and Sustainable Development across the Primary Curriculum* (pp. 37–53). Abingdon: Routledge.

Pimlott-Wilson, H. and Hall, S. (2017) Everyday experiences of economic change: Repositioning geographies of children, youth and families. *Area*, 49(3), pp. 258–265.

Prakoso, S. (2018) Essential qualities of children's favourite places. *IOP Conference Series: Earth and Environmental Science*, 126, p. 012003.

Ridgers, N., Knowles, Z. and Sayers, J. (2012) Encouraging play in the natural environment: A child-focused case study of Forest School. *Children's Geographies*, 10(1), pp. 49–65.

Rieh, S.-Y. (2020) *Creating a Sense of Place in School Environments: How Young Children Construct Place Attachment*. New York: Routledge.

RTÉ [Raidió Teilifís Éireann] (2021) *News2Day May 28th 2021: Central Model School Pupils' Campaign for Reopening of the Sean McDermott Street Pool*. https://twitter.com/news2dayRTE/status/1398304203475402752?s=20

Salameh, F. and Çubukçu, E. (2018) Mental maps of Syrian refugees' children of Syrian and Turkish neighbourhoods. Paper presented at the *Refugees and Forced Immigration/III. International Interdisciplinary Conference on Refugees and Forced Immigration*, 12–13.10.2018, Istanbul, Turkey.

Schlemper, M. B., Stewart, V. C., Shetty, S., and Czajkowski, K. (2018) Including students' geographies in geography education: Spatial narratives, citizen mapping, and social justice. *Theory & Research in Social Education*, 46(4), pp. 603–641.

Scoffham, S. (ed.) (2016) *Teaching Geography Creatively*. Abingdon: Routledge.

Shier, H. (2001) Pathways to participation: Openings, opportunities and obligations. *Children & Society*, 15(2), pp. 107–17.

Soja, E. (1996) *Thirdspace: Journeys to Los Angeles and Other Real-and-Imagined Places*. Oxford: Blackwell.

Stephens, L., Spalding, K., Aslam, H., Scott, H., Ruddick, S., Young, N. and McKeever, P. (2017) Inaccessible childhoods: Evaluating accessibility in homes, schools and neighbourhoods with disabled children. *Children's Geographies*, 15(5), pp. 583–599.

Thew, H., Middlemiss, L., and Paavola, J. (2020) "Youth is not a political position": Exploring justice claims-making in the UN Climate Change Negotiations. *Global Environmental Change*, 61, p. 102036.

Torres-Harding, S., Baber, A., Hilvers, J., Hobbs, N. and Maly, M. (2018) Children as agents of social and community change: Enhancing youth empowerment through participation in a school-based social activism project. *Education, Citizenship and Social Justice*, 13(1), pp. 3–18.

UNICEF (2021) *Education: Every child has a right to learn*. www.unicef.org/education.

Valentine, G. (2004) *Public Space and the Culture of Childhood*. Aldershot: Ashgate.

Waite, S. (2017) *Children Learning Outside the Classroom: From Birth to Eleven*. 2nd edition. London: Sage.

Waller, T., Ärlemalm-Hagsér, E., Hansen Sandseter, E., Lee-Hammond, L., Lekies, K. and Wyver, S. (eds.) (2017) *The Sage Handbook of Outdoor Play and Learning*. London: Sage.

Williams, (2021) Poverty, place and learning. In V. Cooper and N. Holford (eds.) *Exploring Childhood and Youth* (pp. 108–122). Abingdon: Routledge.

Williams-Siegfredsen, J. (2017) *Understanding the Danish Forest School Approach: Early years education in practice*. 2nd edition. Abingdon: Routledge.

Willy, T. (ed.) (2019) *Leading Primary Geography*. Sheffield: Geographical Association.

Wilson, K., Coen., S., Piakoski, A. and Gilliland, J. (2019) Children's perspectives on neighbourhood barriers and enablers to active school travel: A participatory mapping study. *The Canadian Geographer*, 63(1), pp. 112–128.

Yates, E., and Oates, R. (2019) Young children's views on play provision in two local parks: A research project by early childhood studies students and staff. *Childhood*, 26(4), pp. 491–508.

7

STUDENT VOICE, DEMOCRATIC EDUCATION, AND GEOGRAPHY

Reflecting on the findings of a survey of undergraduate geography students

Lauren Hammond and Grace Healy

Introduction

Educational institutions are places to which people travel, and in which they learn, play, access support, forge relationships, build careers, may feel free or oppressed, and encounter sub-cultures. They can also be places of attachment, which people may dream of, or dread, and can work hard for years to (physically or emotionally) move closer to or further away from. Educational institutions shape spaces and places, just as they are simultaneously shaped by them. Ideas, knowledge, skills, and hierarchies that are (re)produced within the social and spatial boundaries of the institution extend into communities (Kramer, 2019). Institutions engage with societal changes and challenges, and academics and students respond to, and theorise, everyday life and (issues in) the world (Harvey, 2013), in turn shaping disciplines and institutional systems and processes.

Yet, it is important to recognise that the relationships between educational institutions and the communities they serve and society more broadly, vary between places and across time-space (Barker et al., 2010), as well as at institutional and individual levels (Salinas-Silva, 2021). Whether those we teach are described as children, young people, or students, it is of critical importance to recognise that they do not passively experience education (Catling and Martin, 2011; Roberts, 2014, 2017), or for that matter, educational spaces (Kearns, 2020) or institutional regulations and networks of power (Oswell, 2013; Hamilton, 2020). It is important that educators engage with students' experiences of, and perspectives on, education as part of their everyday practice to support the co-creation of socially just educational spaces and systems.

As Seixas (1993: p. 132) argues when considering disciplines as communities, the understanding in the academy is that the most junior member of an academic community 'is able to challenge the most senior' and that 'critical comment on

DOI: 10.4324/9781003248538-10

each other's work is expected of all'. This is significant in how we consider the relationships between 'the student', those that teach, and the academic discipline. hooks' (2003: p. 48) reflection that students – 'especially gifted students of color from diverse class backgrounds' – can struggle in, give up hope, or choose to drop out from educational systems structured around maintaining systems of oppression if they do not encounter the support of democratic educators or communities shows the human impact of not truly engaging with the (young) people that we teach. It is significant to note that these systems of oppression may be purposefully structured and enforced, for example, by single-party governments in authoritarian states where 'education aims to ensure adherence to the national regime and coercive compliance with officially sanctioned decrees, laws and expectations' (Jerome and Starkey, 2021: p. 15). They may also be the result of historical, institutional, and everyday injustices, which are then (re)produced in, and/or through, education if they are not challenged.

In this chapter, we draw on a survey of 333 undergraduate geography students studying across the British Isles. The survey was conducted in late 2020 and aimed to critically explore students' perspectives on their geographical education to date. We begin by examining why it is significant to engage with those we teach, considering this at a multitude of scales from the lecture/lesson to the institutional. We draw upon literature which refers to different educational phases to examine how the 'place' of the student has been constructed and represented in schools and universities. This is significant both as the survey had a reflective element – in which respondents considered earlier phases of their geographical education – and also in considering how research and ideas can be shared between educational phases and spaces. Following this, we introduce the research methodology and modes of analysis, before discussing the findings of the research focussing on students' responses to two questions from the survey; 'what did you find most valuable about your geographical education in school?' and 'what did you find least valuable about your geographical education in school?'. We conclude by arguing that through engaging with students' experiences and views, geography education can become more democratic, with students' rights better respected, pedagogy enhanced, and educators becoming more informed about who they teach.

Democratising geography education through engaging with those we teach

We begin this section by considering the level of the lecture or lesson. Here, we examine the opportunities for students to share their experiences and views in, and about, education, also considering how these opportunities are shaped within (and beyond) the institution in which they study.

Many of those who work in educational institutions – including ourselves – have job titles that suggest we share knowledge and ideas; we teach, lecture, or profess, often in a specific discipline or phase. Yet, teaching in both formal and informal contexts is much more than this, and the problematics of what Freire

(1970) terms 'banking education'[1] have long been recognised in terms of both oppression and pedagogical value. As hooks (2010: p. 64) explains when reflecting on her experiences of teaching in the academy 'genuine learning, like love, is always mutual'. Through day-to-day and longer-term choices in curricula and pedagogy it is important for educators to engage with those they teach for two main reasons. First, to support students in meaning-making through connecting with their everyday experiences of the world and prior knowledge (Roberts, 2014); and second, to ensure that students' perspectives, ideas, and values are respected and engaged with in their education.

The opportunity to openly contribute to academic debate in a safe environment in which different perspectives are respected is an important element of what hooks (2003) terms *democratic education*. In this situation, (young) people are empowered in, and through, their education and are respected as citizens rather than conceptualised and treated as citizens in waiting (Starkey et al., 2014). Here, the school/educational institution is conceived of as a 'community of citizens' (Jerome and Starkey, 2021: p. 16), with all members of this community having 'equal entitlement to respect, dignity and rights' (Ibid.). As part of democratic education, teachers/lecturers can challenge systematic and unseen structures of domination through both what they teach and how they teach it (hooks, 2003), with students also engaging in these processes. Such a position emphasises the need for ongoing consideration of the relationships that exist between the student, the educator, and the subject and/or the discipline, in relation to questions of curriculum, pedagogy, and purpose.

Geography is one of 'humanity's big ideas' and a 'core component of a good education' (Bonnett, 2008: p. 7), which supports an individual in making sense of their life and their relationships to the world in which they live. A pivotal element of geographical education is to help students to critically engage with the systems and processes that exist in the world and to support them in becoming more informed about how they can contribute to society and make decisions about how they act in the world today and in their futures (Maude, 2016; Roberts, 2017). These arguments are significant at the level of the individual lesson, seminar, or lecture and with regard to longer-term considerations of the purposes of geographical education, as well as in considering students' experiences of educational spaces.

In considering democratic education, it is also of value to examine the wider structures, processes, and spaces through which educational institutions engage with students' views. For example, institutions may have student councils and advisory boards or encourage students' attendance at meetings; they may also conduct surveys to provide students with opportunities to share their perspectives; and there are also formal inspection processes which often seek students' views, such as Ofsted in the English context (Biddulph, 2011).

These wider systems and processes are significant in supporting students in raising concerns that may not relate directly to, or go beyond, the teaching and learning of specific topic, subject, or phase. This might include active discussion of air pollution in the locality, the quality and upkeep of the institutional environment, and feelings of safety around, and commuting to, campus. Both here, and in the classroom or lecture theatre, 'experience of democracy needs to go hand in hand with conceptual learning and an understanding of the structures and institutions that support democracy in the wider community and society' (Osler, 2010: p. 12). This is to ensure that students are afforded more than tokenistic opportunities to engage with adults/teachers/academics (Lundy, 2007), and – in the case of geography education – geographical debates in both school and in the world beyond the classroom door.

Here, Lundy's (2007) work on children's rights is helpful in further considering how students can be empowered in, and through, geography education. Considering article 12 of the United Nations Convention on the Rights of the Child (UNCRC) (UNCRC, 1989), which focuses on children's voice in different contexts, Lundy (2007: p. 933) argues that voice is not enough and proposes a model in which children are provided with:

Space: Children must be given the opportunity to express a view.
Voice: Children must be facilitated to express their views.
Audience: The view must be listened to.
Influence: The view must be acted upon, as appropriate.

Lundy's model of child participation conveys both children's right to express their views and for their views to be engaged with by others (Jerome and Starkey, 2021). The model also highlights the intersections between voice and other rights as conveyed by the UNCRC, including the right to non-discrimination (Ibid.). Although the model focusses on children, we argue that it is potentially helpful to educators in all phases of education, as they critically reflect on how the views of those they teach are engaged with at both systemic and everyday levels. For example, in considering if, how, and why, the views and experiences of some people may be privileged over, or obfuscated by, those of others. As Osler (2010: p. 56) explains 'democracy needs diversity', and recognising the importance of different perspectives and responding to the needs of those who are or have been marginalised are key to democratic communities and education.

People, including children, can feel frustrated if/when their ideas and perspectives are not engaged with (Osler, 2010). These feelings and frustrations might manifest in educational institutions, and in everyday life (for example, through climate strikes and activism), with there being increasing discourse as to how children and young people might be 'given' what Rousell and Cutter-Mackenzie-Knowles (2020) describe as a 'voice' and a 'hand' in redressing ecological and environmental emergencies in, and through, education. For example, Dunlop and Rushton (2021) have collaborated with young people, teachers, academics, and artists to develop a 'Manifesto for Education for Environmental Sustainability' and as Dunkley (2022) details, citizen science can be used as an eco-pedagogy to encourage participation in geographical research in education.

However, despite being the institution in which most children go to learn about the world through subjects and socialisation, 'most schools lack mechanisms that allow for the full participation of students in decision-making processes' (Osler, 2010: p. 1). The feelings and frustrations that Osler refers to can be seen to be representative of not only the systematic barriers that can, and often do, exist in schools but also cultural barriers, with young people often being constructed as 'lacking competencies, inferior to teachers in the classroom and passive recipients of educational provisions' (Starkey et al., 2014: p. 428). This can be seen to be representative of both institutional cultures and histories (Oswell, 2013) and the gap between the recognition of children's rights in international treaties such as the UNCRC (1989) and governmental commitments to them and what happens in educational practice in schools (Lundy, 2007; I'Anson, 2018). Here it is significant to recognise that whilst the UNCRC has supported and influenced national and global frameworks for children's rights and guided adults' practice and focus, it also has limitations (Urbina-Garcia et al., 2021). For example, it fails to take in to account differences between places (Ibid.).

We now move on to introduce the enquiry, which actively sought student perspectives on their geographical education to date.

Introducing and situating the research

Researchers and teachers working in geography education have previously engaged with children and young people's experiences of both everyday life and their education. For example, Hopwood (2012) examined young people's conceptions of geography in secondary schools, and Biddulph and Adey (2004) examined young people's perceptions of school history and geography at Key Stage 3 (11–14-year-olds). The notion of researching with, and for, children has informed some of this work, with Hammond (2022) using participatory methods drawn from children's geographies to support young people in sharing their experiences and imaginations of London. As part of this process, the young people in the study chose to share their experiences of schooling (Ibid.). In addition, the Young People's Geographies Project (2006–2011) led by Mary Biddulph and Roger Firth engaged academic geographers, teachers, teacher educators, and students together in conversations about children's geographies and the school geography curriculum. This project aimed to develop a curriculum and associated pedagogies that could support students in using their lived geographies to develop their geographical understanding (Biddulph, 2012).

By encouraging undergraduate geography students to reflect on their geographical education to date, our research makes a distinct contribution to these debates. Originally planned as a participatory and arts-based research project with school children, for logistical, ethical, and legal reasons it would not have been appropriate or safe to conduct our research as planned in the midst of the COVID-19 pandemic. We redesigned the project as a survey for undergraduate geography students, which aimed to:

1   Examine student perspectives as to the value of geography to their education/a person's education more broadly;

2   Provide an opportunity for students to share their experiences of, and perspectives on, school and university geography; and

3   Examine students views as to the relationships between school and university geography.

The research was conducted following BERA's (2018) ethical guidelines, and ethical approval was obtained from IOE, UCL's Faculty of Education and Society. The survey was shared by the Royal Geographical Society (RGS) with all Heads of Geography Departments at universities across the British Isles, and in mailings to undergraduate geography societies, RGS Ambassadors and Young Geographers. In addition, the survey was shared directly by the researchers with academics working in both geography and geography education, and the survey was also disseminated via Twitter. The survey was open for responses between 10 October and 19 December 2020. Only current undergraduate students of geography were surveyed. This particular demographic have all had the opportunity and/or chosen to study geography at university. As Oakes and Rawlings Smith (2022: p. 32) explain '18 per cent of A level geography students do not choose university, and of the remaining 82 per cent who do go onto university, only 17% study geography'. Indeed, in England, geography becomes an 'optional' subject after students finish Key Stage 3 when they are 14 years of age. We therefore recognise that there are students' perspectives beyond the scope of this research that may differ from those presented in the chapter and are important to engage with.

The survey had partial responses from 403 students, with 333 students completing the survey or engaging with most of the questions. There were responses from students in 31 universities. There was also a response from a student studying at a university in the United States, which was removed from the data set prior to analysis. Of the universities in the British Isles, 16 were members of the Russell Group and seven were post-1992 institutions. Twenty-three of the institutions were in England, two in Scotland, two in Wales, two in Northern Ireland, and one in the Republic of Ireland. Of the responses, there were over 20 students who responded from seven universities (two in Scotland and five in England), six of these universities were members of the Russell Group. In recognising diverse transitions to university and experiences of school systems and cultures, it is pertinent to recognise that 62 of the 333 respondents were studying for their undergraduate degree in a different country to the country in which they went to school; this included 45 respondents who moved from beyond the British Isles.

Responses to the survey were coded inductively to allow themes to emerge from the data as they were interpreted (Gibbs, 2014). The survey began by collecting demographic data to support data analysis, before posing 15 questions which aimed to examine students' experiences of, and perspectives on, their geographical education to date. More detailed analysis of the full findings of the undergraduate survey is reported in Healy and Hammond (in preparation), including consideration of intersectional identities and more broadly the 'geography of geography' (Brace and Souch, 2020). In this chapter, we focus specifically on students' responses to two questions 'what did you find most valuable about your geographical education in school? and 'what did you find least valuable about your geographical education in school?'. We focus on these questions to support critical reflection on students' overarching perspectives on what is and is not valuable in school geography. As both were open questions, we offered the prompt, 'if you wish, you can link your reflection to a particular phase of education, a qualification, or an area or element of geography (e.g., development or fieldwork)', to stimulate discussion. Out of the 333 students who responded to the survey in detail, 202 responded to the 'most valuable' question,179 responded to the 'least valuable' question, with 175 students responding to both questions.

From the outset, we recognise the nature of this research did not support dialogue or co-production between the researchers and students, and so there are limitations to the enquiry in terms of the areas of focus. The survey was designed using a mixture of open and closed questions. The open questions were aimed at encouraging students to respond on matters which concern them. However, due to the open nature of these questions it is not always clear how representative the perspectives shared are. It is also significant to note that 43% of the survey's respondents were in their first year of university, meaning they experienced disruption to their final year of schooling and examinations due to the COVID-19 pandemic. However, due to the nature of the questions we focus on in this chapter, the pandemic was not mentioned directly in any responses to either question.

The most and least valuable elements of school geography: discussing the perspectives shared by undergraduate geography students

The first phase of the analysis identified 14 themes which represent the most valuable elements of school geography shared by the respondents to the survey. In the second phase of analysis, themes were grouped to identify six overarching themes, with original themes presented as sub-themes. The number and percentage of respondents coded as each of the themes is shown in Table 7.1. Responses were coded as relating to all themes that they connected to, and there are relationships between the themes identified. For example, independent research projects in school may also provide a good grounding for studying geography at undergraduate level and help the student better understand the practical applications of the research in everyday life/the relationships between geographical ideas and theory, and the 'real world'.

The overarching themes presented in Table 7.1 show that the majority of students chose to reflect on aspects of their geographical education that are intrinsically unique to geography rather than those that could relate to other subjects, with the exception of qualifications.

TABLE 7.1 Analysis of responses to the question 'what did you find most valuable about your geographical education in school?'

Theme	Sub-Theme	Respondents	
		Number	Percentage
Teachers and teaching – referring to pedagogy and the influence and passion of teachers.		35	17%
Knowledge and skills which provide a grounding for the future (education, everyday life, and employment) – considering how geographical education can support a person in their life and future.		55	27%
Qualifications – referring to the awards gained from studying geography.		4	2%
Ways of studying geography – referring to students' perspectives on how geography is best taught and studied	Independent research (especially in areas of personal interest)	37	18%
	Fieldwork	41	20%
The discipline of geography – considering how geography is represented, communicated, and understood in different spaces.	Nature of the discipline	45	22%
	Nature of school geography	45	22%
	Specific geographies of interest	33	16%
The power and value of a geographical education – considering the value of geography to an individual, society, and/or the Earth.	Connects to everyday life (understanding the 'real world' and/or practical application)	24	12%
	Affective	25	12%
	To think, and see, from different views	19	9%
	Knowledge of the world (people, places, processes, and systems)	26	13%
	Knowledge of issues and events in the world	13	6%

Note: percentages are rounded to the nearest whole number.

   The first phase of analysis on the least valuable elements of geographical education in schools identified 19 themes, with six overarching themes then identified in the second phase of analysis (shown in Table 7.2). Table 7.2 appears to show that in exploring what was least valuable to them, the undergraduate students often chose to focus on what was not taught or taught in a limited and/or deeply

TABLE 7.2 Analysis of responses to the question 'what did you find least valuable about your geographical education in school?'

Theme	Sub-Theme	Respondents	
		Number	Percentage
Nothing was not valuable/ N/A – A response was provided, which either indicated that the respondent wrote N/A or expressed that nothing about their education in school was not valuable		25	14%
Curriculum – referring to all levels of the curriculum (including, but not limited to, the national, planned, enacted, and experienced curriculum)	Narrow curricula (selective geographies taught and a lack of choice)	62	35%
	Framed by racist/imperialist geographies	3	2%
	Lack of skills (e.g., critical thinking, research literacy, and academic writing)	13	7%
	Specific geographies (not of interest to the respondent)	18	10%
Gaps and borders – between school and university geography and school geography and everyday life	Lack of relevance to everyday life and the world beyond the school gate	20	11%
	Between school and university geography	23	13%
Teachers and teaching – referring to pedagogy, and students' relationships with/ views of their teachers	Fieldwork (limited opportunities and/ or fieldwork that does not develop geographical knowledge or skills)	22	12%
	Outdated theories, ideas, and/or case studies used	11	6%
	Teachers unengaged in and/or with geography or teaching geography	37	21%
	Overreliance on textbooks	6	3%
	Limited criticality/engagement with different perspectives	11	6%
	Limited opportunities for research (especially in an area of choice)	21	12%
Assessment – referring to modes of assessment and the impacts of assessment on teaching, learning, and experiences of geography education		46	26%
Experience of/perspectives on geography education – referring to students' direct experiences of/views on studying geography	Unchallenging	15	8%
	Difficult workload	4	2%
	Repetitive	9	5%
	Unengaged classmates	6	3%

Note: percentages are rounded to the nearest whole number.

problematic way. The perspectives of the respondents as undergraduate students – 93% of whom were between the ages of 18 and 22, and 97% of whom were under 25 – are significant as they were temporarily close to their school education but also close to the frontiers of geography in their university studies. This provides the students with a unique positionality in their critical reflections.

We now move on to discuss the themes in more detail, sharing responses from the survey to illuminate discussion. Here, we focus on three main themes 'teachers and teaching', 'assessment', and the relationships between geography as an academic discipline, school subject, and everyday life. We focus on teachers and teaching as a primary theme that was identified through analysis of responses to both questions, and assessment as a significant number of responses examined the interplay between assessment and teaching. The focus on the relationships between geography in different spaces enables critical examination of students' affective and embodied experiences of education.

One in six (17%) of the respondents valued teachers and teaching as an integral part of geographical education in schools. As one respondent explained:

> I had a teacher who would stay with me for the whole of my GCSEs it was over these years at school I understood the importance of the student teacher relationship.

This respondent can be seen to express the importance of positive student–teacher relationships to students' engagement with geography and their experiences of education. As another respondent shared, 'I found the teachers most valuable, while they didn't know a lot about current geographical issues, they made lessons exciting, were passionate about the subject and made us love the subject before getting bogged down in getting good grades'. Here, the response seems to position the value of learning geography above what might be conceptualised as an extrinsic educational aim of gaining qualifications. The response also suggests that the student feels that their teacher was not particularly engaged with recent geographies. Whilst we do not know the reason for this comment, it can be seen to highlight that students can, and do, value teachers' passion about geography and use of certain pedagogies to engage them in the subject.

Analysis of the responses to the least valuable elements of a geographical education also demonstrate the importance of teachers and teaching. Here, analysis shows that 21% of the respondents felt disappointed with teachers who were unengaged with geography and geography education, 6% with the use of outdated theories, ideas, and case studies, and 3% with an overreliance of textbooks. For example, one respondent reported that the least valuable element of their geographical education was 'learning outdated concepts and theories that have since been disproven or don't fit current knowledge of how the world now operates' and another expressed 'teachers lacked knowledge outside of the textbook. At A level content was rushed and we were only taught to pass an exam'.

Analysis of both questions can be seen to demonstrate frustration with how assessment systems shape the teaching and learning of geography in schools, especially as most respondents focussed on phases of education which conclude with national examinations either implicitly or explicitly in their comments. For example, in response to the question on the most valuable elements of a geographical education, no responses mentioned primary education or education before the General Certificate of Secondary Education (GCSE, England) whereas 14 responses directly noted GCSEs, 20 noted Advanced level (A level, England), ten noted International Baccalaureate (IB), and three noted Highers (Scotland).

One quarter (26%) of the respondents expressed that examinations were the least valuable element of geographical education. As one respondent explained:

> The least valuable thing that's seen as the most valuable for exam boards is the exams. This is because our focus was centred on making sure we finished in the given time (cramming answers) and revising exam techniques for multiple marks question.

The impact of standardised summative assessments on shaping teaching and learning in schools was echoed in many other responses. For example, another student stated 'most of it was just a memory test. They completely lost sight of what education is meant to be. I don't remember everything as the knowledge was being taught solely for the exam, not to educate us in life'. Whilst it is not clear who the 'they' this student mentions are (perhaps a collective they for exam boards, policymakers, and teachers), analysis suggests that students can, and often do, feel that national examinations have negatively shaped their experiences of school geography, including the teaching they have engaged with. Implicit in this response is also the student's perceptions on the purposes of geographical education – which might be conceptualised in this example as providing a person with a better understanding of (their) life in the world.

Analysis also showed that the focus on examinations impacts on the development of academic skills required at university and which may also be of value in everyday life. For example, 6% of respondents reflected that school geography does not always promote criticality and/or engagement with different perspectives. Whilst this is a relatively small percentage, we argue it is worthy of examination as it potentially has a significant impact on the development of skills. As one student commented:

> Geography in school is very structured and limited in the sense that there isn't much opportunity to divert from the course outline and interpret ideas in an original way. There is also very little opportunity to question or criticise the information and theories you are given as a student, you are taught to just accept everything.

However, the negative impacts of examinations on geographical education can be countered by the kinds of approaches to studying geography that students value. As the survey suggests, there are powerful ways of engaging students in geography through teaching. For example, a fifth (18%) of respondents valued undertaking independent research, especially on geographies that interest them, and 20% of respondents reported on the value of fieldwork.

As the following responses reveal, fieldwork and independent research not only develop important geographical skills but support students in connecting theories and concepts with, and to, everyday life. As one student commented, they most valued 'trips around the local area and the UK as they allowed us to develop a wider insight into the world around us, a skill that is incredibly useful at uni'. Another respondent argued that fieldwork also supports students' social development, (potentially) introduces them to hobbies, and helps them to develop skills which are important in both geography and everyday life:

> Fieldwork very important for personal development (first time away from home for some students). Also learning about my local region geography was valuable and made me appreciate where I lived. Also learning basic outdoor skills e.g., using a compass – this led me onto many passions/ hobbies I now have.

The connection to everyday life, both in terms of the practical applications of geography and understanding the world, was something that 12% of students said they valued. For example, one respondent explained the link to 'the outside world and current affairs – the ever-changing nature of geography lessons based on what was happening in real time made lessons interesting'. The affective nature of geography was highlighted by 12% of students. Here, analysis suggests that what is taught and how it is taught (and the interplay between them) has the potential to deeply impact on students' interests in geography as a discipline and geographies in, and of, everyday life. As one student commented:

> I believe that the most valuable thing about geographical education is having the opportunity to learn about other places to your home country. It provides the opportunity to broaden your horizons and learn about social inequalities throughout the world.

With another student stating that geography 'opened my eyes to thinking about things on a variety of temporal and spatial scales', suggesting that their experiences of school geography had changed how they viewed the world. Finally, analysis found that 11% of respondents reported frustrations around the disconnects between everyday life and the school geography, and 13% in relation to the disconnect between the academic discipline and school subject. As one student reflected, 'I felt it wasn't inspiring enough and that it was very rigid the structure of education, set up to pass exams rather than encourage students to learn more about the subject, the world itself and overall make a difference'.

Conclusions

As this chapter has demonstrated, engaging in, and with, geography education is an embodied and affective experience. Through their education, students directly engage with those that teach them (teachers and academics), their peers, how they are taught (pedagogy), what they are taught (geography), and how they are assessed. These experiences shape, and are shaped by, students' relationships with the subject they are studying, those that that teach them, and wider influences on the curriculum (e.g., exam specifications and national curricula), which directly and indirectly shape educational practice and students' experiences. As we considered earlier in the chapter, students' relationships to geographical education are also shaped by educational spaces and students' intersectional identities, values, views, and interests.

Through engaging with undergraduate students' perspectives on their geographical education, this chapter has revealed that students deeply value their geography teachers, and they appreciate their role in inspiring/developing their passion for geography. The chapter has also revealed that students can, and do, feel a sense of frustration at assessment systems shaping teaching and learning in school geography. Here, analysis suggests that respondents perceived that summative national assessments (which, in England, include GCSEs and A levels) are limiting teacher freedoms, encouraging pedagogies of read and remember, reinforcing 'the gap' between school and university geography by limiting criticality and independent research in school geography, and potentially affecting students' relationships to geography.

As a large-scale study of undergraduate students' perspectives on their geographical education to date, our study provides an initial insight into students' views on the geography education taught and experienced in schools. Moving forward, further research is needed to critically examine the relationships between student voice, democratic education, and geography in policy and practice in different educational spaces and phases. As is increasingly examined in education for sustainability (Dunlop and Rushton, 2021; Dunkley, 2022), citizenship education (Jerome and Starkey, 2021) and in consideration of injustices in education and everyday life (Hammond, 2021), respecting children's rights and engaging with their diverse voices is of critical importance to children, education, and society. Here, the intersections between children, education, and geography are significant in further examining: how children and young people experience and view education; if they perceive that geography education empowers them in their everyday lives and futures; and how education (re)produces spaces and the impacts this has on young people and their rights. These endeavours require methodologies and pedagogies of participation and active engagement with young people and those who work in education – including, but not limited to, teachers, academics, and policymakers.

In concluding this chapter, we argue that academic debate and practice can be enhanced in geography education by truly engaging with students. Here, genuine, rather than tokenistic engagement with students' ideas and perspectives can be used to inform pedagogy to support students in building on what they already know (Roberts, 2014) and to bring the subject alive through encouraging students to think about the world in different ways and connecting with everyday geographies (Catling and Martin, 2011). At a more systemic level, engaging with students' experiences and perspectives can help to inform and support democratic education, through enabling educators, researchers, and policymakers to critically reflect upon the human impacts of educational policy and practice.

Note

1 Banking education is a term introduced and made famous by Paulo Freire (1970). As Freire explains 'in the banking concept of education, knowledge is a gift bestowed by those who consider themselves knowledgeable upon those they consider to know nothing' (p. 45). In this situation, students merely receive and store the information bestowed on to them by the educator. This conceptualisation of education may support and maintain systems of oppression. Further, it negates that 'knowledge emerges only through invention and reinvention, through restless, impatient, continuing, hopeful inquiry human beings pursue in the world, with the world, and with each other' (Ibid.).

References

Barker, J. Alldred, P. Watts, M. Dodman, H. (2010) Pupils or prisoners? Institutional geographies and internal exclusion in UK secondary schools. *Area*. 42(3), pp. 378–386.

BERA (2018) *Ethical guidelines for educational research*. Fourth edition. www.bera.ac.uk/publication/ethical-guidelines-for-educational-research-2018-online (Accessed 29.06.2021).

Biddulph, M. (2011) Articulating student voice and facilitating curriculum agency. *Curriculum Journal*. 22(3), pp, 381–399.

Biddulph, M. (2012) Young people's geographies and the school curriculum. *Geography*. 97(3), pp. 155–162.

Biddulph, M. Adey, K. (2004) Pupil perceptions of effective teaching and subject relevance in history and geography at Key Stage 3. *Research in Education*. 71(1), pp. 1–8.

Bonnett, A. (2008) *What is Geography?* London: SAGE Publications.

Brace, S. Souch, C. (2020) *Geography of geography: the evidence base*. Available at: www.rgs.org/geography/key-information-about-geography/geographyofgeography/report/geography-of-geography-report-web.pdf/ (Accessed 12.11.2021).

Catling, S. Martin, F. (2011) Contesting powerful knowledge: The primary geography curriculum as an articulation between academic and children's (ethno-) geographies. *The Curriculum Journal*. 22(3), pp. 317–335.

Dunkley, R. (2022) Looking closely for environmental learning: Citizen science and environmental sustainability education. In L. Hammond. M. Biddulph. S. Catling, J.H. McKendrick (eds) *Children, education and geography: Rethinking intersections*. Abingdon: Routledge.

Dunlop, L. Rushton, E. (2021) 'Why we need a manifesto on education for environmental sustainability'. *BERA blog*. www.bera.ac.uk/blog/why-we-need-a-manifesto-on-education-for-environmental-sustainability (accessed 10.08.2021).

Freire, P. (1970) *Pedagogy of the Oppressed (Translated by Myra Bergman Ramos)*. London: Penguin Books.

Gibbs, G. (2014) Using software in qualitative analysis. In U. Flick (ed.) *The SAGE Handbook of Qualitative Data Analysis*. London. SAGE.

Hamilton, A. (2020) The white unseen: On white supremacy and dangerous entanglements in geography. *Dialogues in Human Geography*. 101(3), pp. 299–303.

Hammond, L. (2021) London, race and territories: Young people's stories of a divided city. *London Review of Education*. 19(1), pp. 1–14. https://doi.org/10.14324/LRE.19.1.14.

Hammond, L. (2022) Recognising and exploring children's geographies in school geography. *Children's Geographies*. 20(1), pp. 64–78 https://doi.org/10.1080/14733285.2021.1913482.

Harvey, D. (2013) *Rebel cities: From the right to the city to urban revolution*. London: Verso.

Healy, G. Hammond, L. (in preparation). *Listening to those we teach: undergraduate students' perspectives on geographical education*.

hooks, b. (2003) *Teaching community: A pedagogy of hope*. Abingdon: Routledge.

hooks, b. (2010) *Teaching critical thinking: Practical Wisdom*. Abingdon: Routledge.

Hopwood, N. (2012) *Geography in secondary schools: Researching pupils' Classroom experiences* London: Bloomsbury Academic

I'Anson, J. (2018) Children's rights. In T.G.K. Bryce. W.M. Humes. D. Gillies. A. Kennedy. (eds) *Scottish education*. Fifth edition. Edinburgh: Edinburgh University Press, pp. 545–553.

Jerome, L. Starkey, H. (2021) *Children's rights in diverse classrooms: Pedagogy, principles and practice*. London: Bloomsbury Academic.

Kearns, G. (2020) Topple the racists 1: Decolonising the space and institutional memory of the university. *Geography*. 105(3), pp. 116–125.

Kramer, C. (2019) Geographies of schooling: An introduction. In H. Jahnke. C. Kramer. P. Meusburger (eds) *Geographies of schooling. Knowledge and Space*. Cham: Springer, pp1–16.

Lundy, L. (2007) 'Voice' in not enough: Conceptualizing article 12 of the United Nations Convention on the Rights of the Child. *British Educational Research Journal*. 33(6), pp. 927–942.

Maude, A. (2016) What might powerful geographical knowledge look like? *Geography*. 101(2), pp. 70–76.

Oakes, S. Rawlings Smith, E. (2022) What constitutes a good a level geography education. *Teaching Geography*. 47(1), pp. 32–35

Osler, A. (2010) *Students' perspectives on schooling*. Maidenhead: Open University Press

Oswell, D. (2013) *The agency of children: From family to global human rights*. Cambridge: Cambridge University Press.

Roberts, M. (2014) Powerful knowledge and geographical education. *Curriculum Journal*. 25(2), pp. 187–209.

Roberts, M. (2017) Geography education is powerful if . . . *Teaching Geography*. 42(1), pp. 6–9.

Rousell, D. Cutter-Mackenzie-Knowles, A. (2020) A systematic review of climate change education: giving children and young people a 'voice' and a 'hand' in redressing climate change, *Children's Geographies*. 18(2), pp. 191–208.

Salinas-Silva, V. (2021) Teachers' territorialities. An expanded definition of teachers' professional practice in rural education. PhD Thesis: University College London.

Seixas, P. (1993) Community of inquiry as a basis for knowledge and learning: The case of history. *American Educational Research Journal*. 30(2), pp. 305–324.

Starkey, H. Akar, B. Jerome, L. Osler, A. (2014) Power, pedagogy and participation: Ethics and pragmatics in research with young people. *Research in Comparative and International Education*. 9(4), pp. 426–440.

UNCRC. (1989) *Convention on the Rights of the Child*. Geneva: United Nations Human Rights.

Urbina-Garcia, A. Jindal-Snape, D. Lindsay, A. Boath, L. Hannah, E.F.S. Barrable, A. Touloumakos, A.K. (2021) Voices of young children aged 3–7 years in educational research: An international systematic literature review. *European Early Childhood Research Journal*. DOI: 10.1080/1350293X.2021.1992466.

8

THE VALUE OF GEOGRAPHY TO AN INDIVIDUAL'S EDUCATION

David Lambert and Kelly León

Introduction

This chapter focuses on the person. Our title periphrastically points us not only to the value of geography in school, or its value to education or to society, it specifically asks to consider the individual and their education. This we are happy to do, for in our view education is a process that centres the person: their individual experiences, the relationships they can forge through their encounters with teachers, fellow students, and new knowledge – all of which requires effort and even struggle. Teachers attempt to build meaningful relationships with students: they try to get inside their minds and find ways to connect, nurture, and sometimes provoke – and the way this happens (and the response it generates) is unique to every young person. For a long time, scholars have used the metaphor of conversation to articulate this process, for education as conversation (Lambert, 1998) is a powerful way to grasp the relational aspects that require both the willingness to listen as well as to tell on both sides (teachers and students).

Geography in the education conversation

It perhaps goes without saying that the conversation in education is, if not unequal, somewhat lopsided. Children and young people are in formal terms the learners, and the adults are the teachers, defining a very particular kind of conversation. It has a particular morality about it – so presumably teachers have to have something they believe is important to teach, something they consider worthwhile (and relevant and enjoyable) for young people to learn. Of course, John Dewey (1916), the philosophical founding father of learning by doing and the encouragement of curiosity and experiential learning was onto this truth over a century ago. Furthermore, it was because of his 'child centredness' (not despite it) that he explained the

DOI: 10.4324/9781003248538-11

crucial role of the teacher as he did – he expended quite some time considering the importance of specialist subject knowledge. The teacher's task is to find ways to enable young people to move beyond the known, the comfort of the familiar, and their own instinctive curiosities or interests. In doing geography, Dewey wrote

> The mind is moved from the monotony of the customary. And while local or home geography is the natural starting point . . . it is an intellectual starting point for moving out into the unknown, not an end in itself.
>
> *(p. 240)*

Dewey specifically recognised the value of geography in supplying the individual with 'subject matter which gives background and outlook, intellectual perspective, to what might otherwise be narrow personal actions' (Ibid., p. 236).

What he also went on to say about the officially sanctioned school subject is even more important to note. One of his abiding themes – and why, in our view, he has frequently been misconstrued as favouring 'doing' and 'experience' over everything else – was his critical take on the nature of the school curriculum as he saw it, consisting solely of 'known facts' clothing the educational experience in the 'second-hand garments' of others. Thus, although he saw both history and geography as the 'information studies par excellence of the schools' (Dewey, 1916, p. 210), he also knew that this carried risk: that school subjects such as these tended to settle comfortably on what simply had become customary to teach and learn. It is worth reading at length what he had to say about geography, specifically in the context of what he saw as its profound purpose – to 'tie together' society and nature for students:

> The classic definition of geography as an account of the earth as the home of man expresses the educational reality. But it is easier to give this definition than it is to present specific geographical subject matter in its vital human bearings. . . . When the ties are broken, geography presents itself as that hodge-podge of unrelated fragments too often found. It appears as a veritable rag-bag of intellectual odds and ends: the height of a mountain here, the course of a river there (etc) . . . the capital of a state. The earth as the home of man is humanizing and unified; the earth viewed as a miscellany of facts is scattering and imaginatively inert.
>
> *(Ibid., p. 240)*

In other words, with a subject like geography, which *in principle* is easy to justify in the school curriculum, we had better be clear about what we are trying to achieve because without a clear sense of direction and purpose, geography can become mundane and, has often been said, more a burden on the memory than a light in the mind (Graves, 1980; Lambert, 2003). But not only this. We need a sound theoretical frame to help us move successfully from the stirring rhetoric of geography as a 'fundamental fascination' and 'one of humanity's big ideas' (Bonnet, 2008: p. 1) to an enacted curriculum that really can provide young people with 'a different view' – that is, with new ways of seeing and understanding (Lambert, 2009; GA, undated).

In this chapter we discuss the enormous power and potential of geography in education for the individual and, crucially, the necessity of curriculum thinking in bringing this potential to reality. We do this from two contrasting international perspectives. First, from a setting in which school geography employs a relatively strong position (England), a nation within the UK, which has long-standing traditions of supportive institutional and policy contexts (for example, geography is a national curriculum subject and a named subject in the English Baccalaureate). And second from California, a US state in which the institutional and policy structures are far less supportive of school geography (for example, a stand-alone course in geography is not a requirement for university admission in California unlike the mandated semesters of US history and civics).

International comparisons are always tricky because the political, economic, cultural, and social contexts are always in flux. We do not therefore seek to draw any general conclusions from this: our comparisons are used simply to illustrate aspects of our discussion. These different perspectives serve to remind us that school curriculum space is universally scarce – but that this challenge plays out in different ways. It is valuable, and it is right to question whose purposes it serves to have a significant proportion of this space occupied by something called geography. Thus, to what extent is geography the keystone of an essentially *conservative* school curriculum, and 'imaginatively inert' as we suspect Dewey was concerned to expose all those years ago? Alternatively, is it possible to reimagine school geography in a form that is more open and dynamic, that can break away from widely held perceptions that it remains traditional or even elitist (Pirbhai-Illich and Martin, 2022)? Finally, if it is unable to free itself of its conservative 'shadow' (Rudolph et al., 2018), on what grounds can, or should, geography resist alternative subject constructions (deemed more dynamic or responsive to students' needs) such as 'Ethnic Studies' (ES) in parts of the United States, including California?

We consider these questions in some detail, particularly examining further the real and perceived inadequacies of school geography, before going on to imagine more appropriate curriculum scenarios for geography in school – for this day and age, including for example, addressing urgent existential challenges of the human epoch (Lambert, 2021; Morgan, 2012). In so doing, we take a significant cue from the GeoCapabilities project which first attempted to apply in practice the nature of what has become known as Future 3 curriculum thinking in geography (Lambert et al., 2021; Mitchell and Béneker, 2022). As we hope to show, one of our intentions is to avoid or even negate the useless 'Punch and Judy' (Claxton, 2021) arguments over simple binaries which plague educational studies: for example, progressive vs traditional; curriculum vs pedagogy; or even inclusive vs exclusively elite. To reiterate: to say that geographical knowledge should be a part of any individual's education is unremarkable. But to judge its value is another matter, as this very much depends not only on what is considered to be 'geographical' but also on the quality of its teaching.

Geography under a shadow

In England

School geography, at least in secondary schools, emerged in the context of the competition between world superpowers at the end of the nineteenth century. In this way, geography contributed to Britain's imperial project, most famously signalled by Sir Halford Mackinder when he called for 'all our teaching (to) be from a British standpoint' (Mackinder, 1911: p. 83). At the same time, geography also played an important domestic role in presenting to children a national space, a 'united kingdom' in which the different landscapes, types of work, and regional identities could all be collected together as elements of a distinctive whole: Britain's geography. Now of course geography, the discipline, does not stand still and so too with school geography. How place, society, and space are understood changes, and the way young people are asked to study Britain in the world at school has also changed. However, although crude and racist environmental determinism, the central element defining the early twentieth-century 'British standpoint', has long been discredited, it has cast a shadow (Kearns, 2020, 2021). To this day, the subject is unusually 'white' as Standish (2021) has recently pointed out (p. 142), and some say even structurally racist (Puttick and Murrey, 2020). Certainly, the question of who studies whom, and through what intellectual lenses, is still one that geography – as a discipline and as a school subject – continues to struggle with.[1]

In broad terms, progressive educational thought in England has mixed a Deweyian child-centred pedagogical approach with a radical attitude to curriculum reform. The latter was influenced by the new sociology of education of the 1970s (Young, 1971) whose analysis saw traditional subjects like geography as part of the problem – underpinning a school curriculum serving the interests of the elite through its focus on the 'knowledge of the powerful'. Such progressive thought was manifest in geography education in England, through a growing interest not in the curriculum *per se* but in 'learning': first, through Frances Slater's innovative contributions (Slater, 1982) and more recently by Margaret Roberts's work on 'enquiry' (Roberts, 2013). Both these scholars shared an interest in the geography – but in others this was far less apparent, as in David Leat's *Think Through Geography* (Leat, 1998) in which geographical knowledge seemed almost incidental to the process of education – as if 'learning' itself was the only goal. It isn't of course. As we know from Bruner (1963) and many others, the object of study – *what* is learned – is of great importance. Nevertheless, progressive thought that sought to undermine the hegemony of 'elitist' subjects has turned out to chime quite well with the neoliberal educational gaze of more recent times, which has fixated on flexibility, generic competence, and learning to learn – dazzled by the thought of schools 'building learning power' ultimately to meet the needs of late-stage capitalism (Mitchell, 2020).

It was to resist, or at least re-balance, the anti-subject *zeitgeist* that motivated the Geographical Association's (GA's) *Manifesto* (GA, undated) with its strong statement of purpose to change young minds, captured through its title *A Different View* (Lambert, 2009). On a grander scale altogether (and at about the same time), the knowledge-blind tendency in school curriculum matters was also singularly addressed by Michael Young in his famous *volte face* and the publication of *Bringing Knowledge Back In* (Young, 2008). In England then, school geography at least had the chance to climb from beneath its own shadow in the last decade or so, by proposing an alternative progressive agenda, one focussed on curriculum questions and not solely on 'emancipatory' pedagogies. This has been attempted through explorations of Young's social realism and its knowingly playful use of 'powerful knowledge' as a means to reignite interest in the emancipatory power of specialist subject knowledge (for example, see Enser, 2021; Lambert, 2018; Maude, 2016; Morgan and Lambert, 2020). This interest has been taken up in geography education internationally, partly through the GeoCapabilities project, which we shall return to later (Bustin, 2019; Lambert et al., 2015; Lambert et al., 2021).

In California

Although education in the United States is predominantly a state rather than federal matter, by the end of the nineteenth century the role of geography in education was largely unquestioned nationally. Geography was thought to be essential in the education of the children of immigrants populating the young, turbulent, and expanding nation in the nineteenth century (McDougall, 2015). Its role was to present, in effect, a 'united states' – an immeasurably more difficult task than a 'united kingdom' given the territorial scale and diversity of the United States. But school geography has not served this purpose for some time. Ever since the 1916 National Education Association report (Dunn, 1916), which recommended that history (and especially American history) should be taught from grades 7 to 12, geography has been judged to be 'incidental to history' (see also Bonnett, 2008: p. 51) and undeserving of precious curriculum space on its own account. In the process of modernising the curriculum for the twentieth century, geography in US schools was therefore demoted. According to contemporary analysis (when of course John Dewey's influence was present), this was in large part a result of the perceived low-level intellectual calibre of the subject. For instance:

> Leaders of school geography are in large measure to blame. . . . The books appear filled with a heterogeneous aggregate of facts about the earth . . . There appears only a dishing up of a great number of facts of every sort. . . . School geography . . . must give a rational account of itself and a justification for its retention or it stands in danger of wholly or in large part disappearing.
> *(Fairbanks, 1927: p. 14; cited in McDougall, 2015: p. 51–52)*

This was a stinging assessment of the subject's complacency, and higher education was not exempt. Thus, famously in 1948, the President of Harvard declared geography 'not a university subject' and disbanded the department. Doubtless, there were political and economic contextual circumstances that complexify this rather high-handed judgement, but the damage to the discipline and geography's identity in the secondary curriculum was immense.

Today, California is one of only nine US states that still requires geography in high school. However, as we have seen, the state-mandated geography standards are in any case embedded (often buried) in predominantly history-focused social studies courses. Whether or not a student takes a recognisable, stand-alone geography course varies by school district and, even in California, the great majority of students will never have that opportunity (California Department of Education, 2019). Indeed, geography is vulnerable when forced into competition with other contenders for scarce curricular space. While in the English context, geography has seen off its upstart rivals (rural studies, world studies, integrated humanities etc.) owing to its institutional strength (Goodson, 1993), in California this has not been the case.

Given the shadow of geography's imperial and nation-building past, one fascinating example of a 'new subject' gaining popularity as a school subject today is Ethnic Studies (ES). This is an interdisciplinary domain, widely offered in US higher education institutions (including Harvard), that centres on the history, experiences, and struggles for liberation of marginalised groups in the United States – matters which, advocates argue, are not afforded enough curricular space in traditional courses – including in high school. While geography's weak curricular position in California wanes further, the number of public schools and districts offering or requiring ES courses is rapidly increasing (Buenavista, 2016; Cuauhtin et al., 2019). This at least raises the question of whether the expansion of ES as a research discipline and its increasing popularity as a school subject are the result of established school subjects like geography disregarding marginalised groups and societal power dynamics in the curriculum content selections. More prosaically, if the rise of ES finally saw the end of recognisable courses in geography in California schools, would anyone notice? Would anyone care, and if so, why?

For those with an understanding of geography's educational potential, it would appear that school geography advocates in the United States (there is a National Council for Geographic Education, linked with the American Association of Geographers, not to mention the enormously wealthy National Geographic) have failed in large part to convey the significance of geographical knowledge perspectives to the general public, education policymakers, and to school district officials who have the authority to decide its fate. This is despite successive reports on the inadequacy of geographical 'proficiency' (United States Government Accountability Office, 2015) and despite the celebratory tones that have accompanied the publication of successive editions of *Geography for Life*, the national standards for geography (Heffron and Downs, 2012). It appears that an old, established school subject with fractured links to the wider discipline (geography) is being replaced. A new subject which claims strong higher education links as well as emancipatory educational potential is a clear replacement candidate. Thus, California's governor signed into law the nation's first model ES curriculum, and more recently mandated an ES course as a high school graduation requirement (Fensterwald, 2021). Academics, teachers, and grassroots organisers have made the argument that California's students, the large majority of whom are students of colour (California Department of Education, 2021a; see also Noguera, 2016), not only need ES but also will benefit from it (Cortés, 2020; Dee and Penner, 2016; Sleeter, 2011; Sleeter & Zavala, 2020). It seems that the advocates of school geography, on the other hand, have been unable to make similar arguments or re-envision it with new light and energy: that is, to take geography out from under its own 'shadowy' past. And of course, as in England, there are conservative people who question whether anything other than geography-as-given should even be contemplated, and so while some states have embraced ES, other more conservative states like Arizona have fought against it (see Cabrera et al., 2014).

Summary

Although this section can be read simply as an account of endless, essentially bureaucratic curriculum turf wars, we believe there is something more significant at play. For at the heart of this story is a concern for the value of education to individuals from diverse social, cultural, and economic circumstances. The uncomfortable truth to emerge from our account is that geography's educational value to the individual is not to be taken for granted. Indeed, there are circumstances in which its value can be deemed to be of little worth, and in California, this sentiment would seem now to hold sway. As the apparently *expendable* occupant of scarce curriculum space, geography appears to be making way for ES. It remains to be seen whether and how ES practitioners rise to the challenge, for as Dewey reminded us, good intentions remain inert and useless if they cannot be translated into sound worthwhile practical teaching.

Geography, arguably, has a broader base and a wider range of purposes than ES, implying something could be lost if geography were to disappear altogether. However, with Dewey, this is perhaps easier to state in principle than it is to demonstrate in practice. It is time to think again about what is 'customary' in school geography. Rising interest in reimagining geography's 'powerful knowledge' may suggest a fruitful way ahead. As the GeoCapabilities project has proposed, this means teachers adopting their curriculum making responsibilities in so-called future 3 scenarios.

Rediscovering the shine

In his textbook on curriculum planning and implementation in school geography (in the UK context), David Gardner draws on Michael Young's 'three futures' heuristic (Young and Muller, 2010), which envisions alternative curricular scenarios, advocating that teachers work towards Future 3. He explains:

> The numbering is slightly misleading, because all three possible 'futures' can exist side by side, but one way to grasp Future 3 (which is the scenario

based on powerful knowledge) is to imagine that it fuses or merges aspects of Future 1 and Future 2.

We can characterise the alternative curricular scenarios as:

F1. Traditional: knowledge-led
F2. Progressive: generic/skills-led
F3. Progressive: knowledge-led.

(Gardner, 2021: p. 27)

We do not have the space to discuss three futures thinking in any detail, save to say that as a device to analyse the form and the function geography curriculum as planned and enacted in schools, it has proved very useful (Bustin, 2019; Enser, 2021; Lambert, 2017a, 2017b, 2018). The contrast between Future 1 and Future 2 scenarios is evident in the previous section – in both England and California. Nevertheless, enacting the curriculum in a Future 3 scenario is challenging and remains elusive. It is not quite as simple as merging scenarios 1 and 2. For a start, it depends on the agency of confident and well-prepared teachers fully accepting their role as the curriculum makers (Gardner, 2021: p. 24), and the key to this is to recognise the significance of geography as powerful knowledge (as distinct from geography seen solely as an expression of knowledge *of the* powerful).

Powerful knowledge can be characterised as dynamic and contestable but also robust and dependable. It is often theoretical and abstract, acquiring systematicity, linking it to bigger powerful ideas and generalisations. It is therefore distinguishable from 'facts', especially knowledge that becomes 'given' or static (or to use Dewey's word, 'customary') – which often happens to the knowledge selections in Future 1 scenarios. But crucially, because powerful knowledge is produced in specialised disciplines or communities of practice it is not arbitrary or fluid, as is usually the case in Future 2 scenarios. A curriculum based on powerful knowledge (Future 3) places stress on how real, material but emergent knowledge is produced and continually develops within specialist communities with shared rules and norms. Thus, the main ambition of Future 3 teaching is to work out in practical terms how all students can be given epistemic access to deep and critical disciplined thought centred on a definable domain or object of study (such as geography whose object of study is the Earth as the home of humankind). In other words, how can all students gain access to the 'power' of abstract and theoretical knowledge that provides them with the capability to think in new ways, to make distinctions between phenomena, to evaluate evidence, to argue the truth of a matter . . . within a specialist domain such as geography (which has the added advantage of being intrinsically inter-disciplinary).

It perhaps goes without saying that there are some assumptions implied here. For instance, it would be naïve to assume that all students 'buy in' to the idea that abstract and theoretical knowledge in school subject specialist domains is in any way 'powerful' to them! Such knowledge is often counter-intuitive and is only even partly grasped after some hard mental effort and a willingness to 'let go' of the immediate or 'common sense'. In other words, to many students such a curriculum may well be resisted because it appears 'alien'. We would just say that it is here where professional teachers must seek ways to convince young people of its merits, and this is done through the curricular decisions they make. Just because students find it challenging is not a good enough reason not to try, for students who do not gain access to powerful knowledge not only are denied access to certain aspects of society – in education, in leisure activities, and in the employment market – but are also condemned to a form of 'capabilities deprivation' (Bustin, 2019: ch. 4), for instance, in their ability to think critically and holistically about society–environment relations. To recall the metaphor used at the beginning of this chapter, their role in wider societal 'conversations' diminish. It is part of the teacher's job to empower and encourage young people to participate. Not just to have an opinion but to engage with knowledge claims in wider society and in their local communities.

It is not surprising therefore that advocates of Future 3, such as in the Geographical Association and the GeoCapabilities project (Bustin, 2019; Mitchell and Béneker, 2022), focus very strongly on the notion of 'thinking geographically'. To enable young people to think geographically 'enables a unique way of seeing the world, of understanding complex problems and thinking about inter-connections at a variety of scales from the global to the local' (Jackson, 2006). It is evidently possible to spell out in some detail what the constituent parts of thinking geographically are (GA, 2012; Jones, 2017). This is not a curriculum, of course, but a framework for helping envision the educational purposes and potential outcomes of geography education, which could help return the 'shine' to geography as a school subject.

A Californian vignette

In the Californian context, school geography is under pressure mainly because the subject in its customary form is perceived to be the quintessentially Future 1 subject: given, information heavy and inert. Such a perception can be reinforced through the manner in which the subject is presented: for example, when Standish (2021: all emphasis in the original) claims that geography is valuable because it '*introduces the world* to the child' (p. 150), inspires 'a sense of awe and wonder' (p. 151), and 'teaches children about *humanity* and will *help them to find their place* within it' (p. 151), he may be unintentionally suggesting a Future 1, one-way street. But young people are naturally curious – and aspects of the world have already come to them through their daily lives as well as via various devices and screens (Walshe and Healy, 2021). Even those who have limited exposure to geography come into high school with interests, concerns, and wonders of their own. Furthermore, these young people are *already* of the world and even those who have travelled only a few miles beyond their neighbourhood have much to say about what they *do* know. The rise of ES, where it has been accommodated in the curriculum by occupying space vacated by geography, is in part related to geography's perceived inability to recognise the humanity of young people themselves or to engage with knowledge they see as more immediately significant for their futures. Certainly, in the case of California schools, ES overtly aims to recognise the lives and experiences of the students themselves. Using the three futures heuristic, ES is a progressive response to the Future 1 inadequacies of a school curriculum considered by many to be obsolete. But how might geography itself respond?

First and foremost, taking GeoCapabilities as our cue, we take as unequivocal that in thinking about the central curriculum question (*what* to teach?), there are in fact two *prior* questions. One of these is to keep in mind *why* teach geography in any case: this question ensures we keep in mind the subject's goals and purposes. But the prime question concerns the students we teach: *who* are they?[2] Put together, with appropriate prioritising, these questions may be enough to reshape Future 1 geography into a Future 3 scenario.

In thinking this through in the Californian context, we felt we needed to at least consult young people. Properly designed evaluative studies of ES will emerge in due course, and these will surely seek fully to research and represent student voices. For the purposes of the present discussion of geography, however, we merely take an opportunistic moment to hear what students say about the developing geography curriculum experienced in Kelly León's classroom in the Sweetwater Union High School District (SUHSD). This is a large, secondary school district located between San Diego and the California/Mexico border, where grade 9 students are still required to take a geography course to graduate high school. The voices reported in the following pages are based on written coursework submissions of one class of students. All names have been anonymised, and the students themselves were aware that their writings were being used and discussed for the purposes of curriculum evaluation and development.

The course was redeveloped in collaboration with the California Geographic Alliance, and in a way, it explicitly takes into account the concerns, interests, and geographies of the students within the district. The upgraded course therefore includes a six-week 'geo-inquiry project' where students take ownership over a self-generated question and the subsequent research. During the 2020–2021 (virtual) school year, students pursued geo-enquiry questions and used ESRI's ArcGIS and StoryMaps to communicate their thinking and research. Students' questions were typically both personal and practical and frequently sought to examine social justice concerns, for example: how does an area's income affect the available skateboarding opportunities in San Diego County? How and where will climate change impact San Diego and what is being done to mitigate it? Where have hate crimes targeting Asian Americans happened in California? Thus, students contemplated issues of personal significance and were taught to use geographic thinking to accomplish this.

For example, when Marco was asked why he wanted to study the relationship between neighbourhood income and school outcomes, he said, 'I really care about educational attainment, not just in my district, but across the country. Being a son of parents that didn't finish high school or didn't even make it to high school, this hit close to home'. Marco was able to use his map to visualise patterns and find connections between income and educational attainment. He claimed that 'geographic language' helped him talk about relationships he never knew existed and that the project encouraged him to think of himself as a member of the community.

Before the course redesign in SUHSD, the geography curriculum was an outdated, uninviting, and static representation of the world 'out there' with little connection to the concerns of students. In much the same way that English Language Arts teachers speak of texts as windows, doors, and mirrors for students (Bishop, 1990), the updated SUHSD geography curriculum allows students to study the world in ways they have never considered, acting as doors and windows to places, concepts, processes, and ways of thinking they did not know existed. Other times, the course becomes deeply personal for students, acting as a mirror and reflecting back students' own lives, values, beliefs, and experiences. It is not that the course asks students to study their communities and themselves as they know them *already* but instead to examine aspects of their lives in new ways. This incentivises teachers to *acknowledge* the diversity of students' individual experiences and imaginations and to use the class as a mechanism for empowering young people to understand aspects of their everyday lives, including the different injustices and inequalities they face.

Geographers investigate these matters, and the discipline has produced valuable insights on place-making and the racialisation of space, society, and economy. Geography teachers in SUHSD have chosen to develop such powerful knowledge with young people and thus speak to some of the silences that ES seeks to address in the school curriculum. And, of course, geography offers more besides – because there is a world 'out there' and particular ways of studying it, the main point that Standish (2021) was attempting to make earlier. It is not so much as introducing the world to the student but introducing ways of seeing it, studying it, and thinking about it – as an object of thought. Again, learning how to think geographically becomes the broad outcome of a Future 3 curriculum scenario.

It would appear from our opportunistic evidence that students who have taken the SUHSD course are able to address the ways school geography contributes to their education. Teresa articulates her biggest takeaway as follows: 'Taking this geography class has actually helped me ask better questions, maybe even more than it has given me answers'. Nico also recognises how taking a geography class has impacted his thinking.

> Learning to think geographically has made me take a different outlook on the world . . . It has helped me broaden my thoughts. Take (people) crossing the border as an example. Many people see that as a threat to them because of things they have heard . . . but thinking geographically can expand your thoughts and make you realise why people make those decisions.

Iliana adds

> Before [taking] this class, I had one idea of geography, which wasn't at all
> what it ended up being. I thought it was just learning about maps, but
> I never realized it involved *us*, the environment, economics, education, and
> our quality of life.

The testimonies from these students provide evidence that they have both learnt
something about the nature of geography (Iliana) and about their own intellectual
growth achieved through the study of geography (Teresa and Nico).

Geography in this day and age

For Alastair Bonnett (2008), geography has always been ultimately concerned with
human survival. Where you build your settlements (and what with?), what you
choose to grow (and where?), where you dispose of waste (and with what consequences?), what you make, and who you trade with (and why?) . . . these are all
matters that benefit from geographical thinking. They are questions that demand
knowledge of both the physical environment (weather and climate, life on earth
and landscape) and human behaviours (in cultural, economic, social, and political realms), for geographical questions are usually concerned with how various
influences and factors interrelate and interact. Thinking geographically is, as Peter
Jackson (2006) explained, *relational*. Thus, although we can talk about 'human' and
'physical' geography – and frequently geographers self-identify as being one or the
other – in truth, *thinking geographically* about an ever-changing world always carries
a relational component which seeks to cross this 'human/society-nature' divide.
Survival is a theme that has been developed explicitly in the context of the school
curriculum (e.g., Lambert, 2013) but, given the mounting existential challenges
that confront humankind in the Anthropocene, there is surely further work to be
done on identifying the most appropriate educational responses.

An appropriate educational response to the human epoch needs to value geographical thinking: thinking that does not put human beings above (or even separate from) nature; that puts locales and nations into their global context; and
which always seeks to understand interconnections. School geography, re-visioned
to enhance young people's intellectual capabilities with regard to thinking *relation-
ally*, not least (but not only) about society–nature relationships and environmental
futures, represents a profound educational response to the challenges of the human
epoch. This is no less than a call for a Future 3 geography curriculum based on
powerful geographical knowledge. Although driven by a concern for the interests
of young people, the subject is a means to help them understand beyond their
immediate experiences and think deeply about a world, where (for example):

• enduring injustices exist not least those that have followed colonial and imperial violence;

- the climate emergency is causing death, economic mayhem, and displacing tens of millions of people across the world;
- biodiversity loss, again on a global scale, is already looking cataclysmic; and
- human–nature relations are now so mixed up, partly a result of almost eight billion people currently living on the planet, that environmental and biological issues (including so-called natural disasters) pose existential threats for millions of people: the COVID-19 pandemic is a spectacular example.

These are real existential threats rather than abstract 'world problems'. They are not unconnected, and it is possible that the pedagogic means need to be found to help students think critically about the unspoken ideological assumptions that underpin the 'there is no alternative' (TINA) approach to economic growth and progress.

In conclusion

In redesigning SUHSD's geography course in Southern California, teachers' insistence, that the revised geography curriculum needed to incentivise students to invest in the course and engage meaningfully with its knowledge content, has required them to think deeply about the curriculum they enact with students. For some, the opportunity to select and engage with content more freely in line with the course objectives is a liberating prospect, but for many it is a challenge, especially given the fact that few teachers in California have academic backgrounds in geography. This lack of subject expertise has not stopped teachers from being thoughtful curriculum makers, but the Future 3 scenario means that the need intentionally to build teachers' disciplinary knowledge is a priority. We acknowledge that exactly the same position could be posited for ES, although in this case many advocates argue that ES should in any case be 'infused' through the curriculum rather than be taught as a separate subject – that insights from ES should be a concern for all teachers. Certainly, there are lessons in the 900-page *Ethnic Studies Model Curriculum* (California Department of Education, 2021b) that could be adopted in geography, such as 'Chinese Railroad Workers' and 'Housing Inequalities: Redlining and Racial Housing Covenants'. Geographical studies of these topics in school can deepen students' understanding of the contemporary United States. The vibrancy and the cultural responsiveness (Gay, 2000) of geography and geography teachers may yet play a part in how this particular argument plays out in various school settings in California.

However, this is always the case, and 'twas ever thus. In England, where secondary school teachers of geography usually have a degree-level qualification in the subject, there is still the need to think anew about the educational purposes of geography in the school curriculum. Indeed, with a continuing and mounting teacher recruitment crisis of specialist geography-trained graduates into teaching in England, there is no need for complacency: 'non-specialist' geography teaching may become more commonplace in England's secondary school. Thus, curriculum-thinking that forces teachers to contend with important questions like *why* they should teach geography and *what* geography they should teach, and which requires mental effort to shape practical decision-making in accordance with disciplinary principles and thinking should arguably become a central pillar in teacher education.

Both these questions are shaped by a profound concern for the individual and their capabilities to engage with geography's powerful knowledge. And as we have argued they must be preceded by the *who* question: who are the young people we teach? The challenge of Future 3 thinking is not only to clarify the powerful knowledge in geography (including knowledge that young people are unlikely to encounter in their everyday lives) but also to find ways to engage them with this in meaningful ways.

Notes

1   In this broad context, it is perhaps salutary to note that since the introduction of the national curriculum (for England) by the Thatcher government in 1988, school geography in the UK context has tended to find a more secure place in the curriculum during periods of conservative governments. This alone can explain to an extent why geography is frequently perceived as 'traditional' and serving the interests of the state – and the progressive educator's instinct to do away with it, or (as happened under new labour in the early twenty-first century) have the subject serve the interests of other agendas – such as citizenship or sustainability.
2   The GeoCapabilities project devised some training materials on 'curriculum making' or enactment which outlines this point in a bit more detail:

www.geocapabilities.org/training-materials/module-2-curriculum-making-by-teachers/getting-started/

References

Bishop, R.S. (1990) Windows, mirrors, and sliding glass doors. *Perspectives*, *6*(3) ppix-xi.
Bonnet, A. (2008) *What is Geography?* London: Sage.
Bruner, J. (1963) *The Process of Education* (Vintage Edition). New York: Vintage Books/ Random House.
Buenavista, T.L. (2016) The making of a movement: Ethnic studies in a K-12 context. In Sandoval, D. M., Ratcliff, A. J., Buenavista, T. L., & Marín, J. R. (Eds.), *"White" Washing American Education: The New Culture Wars in Ethnic Studies* (pp. vii-xxviii). ABC-CLIO.
Bustin, R. (2019) *Geography Education's Potential and the Capability Approach*. Cham: Palgrave Macmillan.
Cabrera, N. L., Milem, J. F., Jaquette, O., & Marx, R. W. (2014) Missing the (student achievement) forest for all the (political) trees: Empiricism and the Mexican American studies controversy in Tucson. *American Educational Research Journal*, *51*(6), pp. 1084–1118
California Department of Education (2019) *California Basic Educational Data System [Data Set]*. http://www3.cde.ca.gov/download/dq/CourseEnrollment18.zip
California Department of Education (2021a). *Fingertip Facts on Education in California*. www.cde.ca.gov/ds/ad/ceffingertipfacts.asp (April 29)
California Department of Education (2021b) *Ethnic Studies Model Curriculum (approved by the State Board of Education* (18 March 2021) www.cde.ca.gov/ci/cr/cf/esmc.asp?campaign_id=49&emc=edit_ca_20211014&instance_id=42840&nl=california-today&regi_

id=91238862&segment_id=71635&te=1&user_id=e1859e0bf16d0bf3860420cf7eae3 dfe

Claxton, G. (2021) *The Future of Teaching: And the Myths that Hold it Back*. Abingdon: Routledge.

Cortés, C. E. (2020) *High School Ethnic Studies Graduation Requirement* (Doctoral dissertation, University of California, Riverside).

Cuauhtin, R. T., Zavala, M., Sleeter, C. E., & Au, W. (Eds.). (2019) *Rethinking Ethnic Studies*. Milwaukee, WI: Rethinking Schools.

Dee, T., & Penner, E. (2016) The causal effects of cultural relevance: Evidence from an ethnic studies curriculum. CEPA Working Paper No. 16–01. *Stanford Center for Education Policy*

Dewey, J. (1916) *Democracy and Education: An Introduction to the Philosophy of Education*. New York: MacMillan. (www.fulltextarchive.com/page/Democracy-and-Education/)

Dunn, A. W. (1916) *The Social Studies in Secondary Education: Report of the Committee on Social Studies of the Commission on the Reorganization of Secondary Education of the National Education Association. Bulletin 28*. Washington DC: Government Printing Office. https://files. eric.ed.gov/fulltext/ED542444.pdf

Enser, M. (2021) *Powerful Geography" A Curriculum with Purpose in Practice*. Camarthen: Crown House Publishing.

Fairbanks, H. (1927) *Real Geography and its Place in the Schools*. San Francisco, CA: Harr Wagner.

Fensterwald, J. (2021, October 8). California becomes first state to require ethnic studies in high school. *EdSource*. https://edsource.org/2021/california-becomes-first-state-to-require-ethnic-studies-in-high-school/662219

GA (2012) *Thinking Geographically*. www.geography.org.uk/write/MediaUploads/ Support%20and%20guidance/GA_GINCConsultation_ThinkingGeographically_ NC_2012.pdf

GA (undated) *GA Manifesto for Geography*. www.geography.org.uk/GA-Manifesto-for-geography

Gardner, D. (2021) *Planning Your Coherent 11–16 Geography Curriculum: A Design Toolkit*. Sheffield: Geographical Association.

Gay, G. (2000) *Culturally Responsive Teaching: Theory, Research, and Practice*. New York: Teachers College Press.

Goodson, I. (1993) *School Subjects and Curriculum Change*. London: Routledge.

Graves N.J. (1980) *Geography in Education*. 2nd edition. London: Heinemann Educational Books.

Heffron, S., & Downs, R. (Eds) (2012) *Geography for Life: National Geography Standards*. Geography Education National Implementation Project (GENIP). https://ncge.org/ teacher-resources/national-geography-standards/

Jackson, P. (2006) Thinking Geographically. *Geography*, *91*(3) pp199–204. See www.geography.org.uk/teaching-resources/videos/thinking

Jones, M. (Ed) (2017) *The Handbook of Secondary Geography*. (Chapters 1 and 2). Sheffield: Geographical Association.

Kearns, G. (2020) Topple the Racists 1: Decolonising the space and institutional memory of the university. *Geography*, *105*(3) pp116–125.

Kearns, G. (2021) Topple the Racists 2: Decolonising the space and institutional memory of the university. *Geography*, *106*(1) pp4–15.

Lambert, D. (1998) Valuing conversation: Opening up the concept of prejudice. *International Research in Geographical and Environmental Education*, 7(2), pp. 146–150.

Lambert, D. (2003) Geography: A burden on the memory or a light in the mind? *Geography*, *88*(1) pp47.

Lambert, D. (2009) A different view. *Geography*, *94*(2) pp119–125.

Lambert, D. (2013) Geography in schools and a curriculum of survival. *Theory and Research in Education*, *11*(1) pp85–98.

Lambert, D. (2017a) The relevance of geography for citizenship education. In Leite, L., Dourado, L., Afonso, A. S. & Morgado, S. (Eds.), *Contextualizing Teaching to Improve Learning: The Case of Science and Geography*. New York: Nova Science Publishers.

Lambert, D. (2017b) Powerful disciplinary knowledge and curriculum futures. In Pyyry, N., Tainio, L., Juuti, K., Vasquez, R. & Paananen, M. (Eds.), *Changing Subjects Changing Pedagogies: Diversities in School and Education*. Helsinki: Studies in Subject Didactic 13. Finnish Research Association for Subject Didactics. https://helda.helsinki.fi/bitstream/handle/10138/231202/Ad_tutkimuksia_13_verkkojulkaisu.pdf?sequence=1&isAllowed=y

Lambert, D. (2018) The Road to Future 3: The case of geography. In Guile, D., Lambert, D., & Reiss, M. (Eds.), *Sociology, Curriculum Studies and Professional Knowledge: new perspectives on the work of Michael Young*. Abingdon: Routledge.

Lambert, D. (2021) "The geography of it all", *Public History Weekly* (18.02.2021) https://public-history-weekly.degruyter.com/9-2021-1/geography-anthropocene/

Lambert, D., Béneker, T. and Bladh, G. (2021) The challenge of 'recontextualisation' and Future 3 curriculum scenarios: an overview. In Fargher, M., Mitchell, D., & Till, E. (Eds.), *Recontextualising Geography in Education*. Cham: Springer

Lambert, D., Solem, M., & Tani, S. (2015) Achieving human potential through geography education: A capabilities approach to curriculum making in schools. *Annals of the Association of American Geographers*, 105(4) pp723–735.

Leat, D. (1998) *Thinking through Geography*. Cambridge: Chris Kington Publishing.

Mackinder, H. (1911) The teaching of geography from an imperial point of view and the use which could and should be made of visual instruction, *The Geographical Teacher*, 6, 83.

Maude, A. (2016) What might powerful geographical knowledge look like? *Geography*, 101(2) pp70–76.

McDougall, W. (2015) Geography, history and true education. *Research in Geographic Education*, 17(2) pp10–89.

Mitchell, D. (2020) *Hyper-socialised: How Teachers Enact the Geography Curriculum in Late Capitalism*. Abingdon: Routledge.

Mitchell, D., & Béneker, T. (2022) Expanding students' concept of 'home': Teaching migration with a geographic capabilities approach. In Hammond, L., Biddulph, M., Catling, S., & McKendrick, J. H. (Eds.), *Children, Education and Geography: Rethinking Intersections*. Abingdon: Routledge.

Morgan, J. (2012) *Teaching Geography as if the Planet Matters*. Abingdon: Routledge.

Morgan, J., & Lambert, D. (2020) For knowledge – but what knowledge? Confronting social realism's curriculum problem. In Barrett, B., Hoadley, U., & Morgan, J. (Eds.), *Knowledge, Curriculum and Equity*. Abingdon: Routledge.

Noguera, P. A. (2016) Race, education, and the pursuit of equity in the twenty-first century. In P. Nogera, J.C. Pierce, & R. Ahram (Eds.) *Race, equity, and education* (pp3–23). Cham: Springer.

Pirbhai-Illich, F., & Martin, F. (2022) De/colonising the geography curriculum, in Hammond, L., Biddulph, M., Catling, S., McKendrick, J. H. (Eds.), *Children, Education and Geography: Rethinking Intersections*. Abingdon: Routledge.

Puttick, S. and Murrey, A. (2020) Confronting the deafening silence on race in geography education in England: Learning from anti-racist, decolonial and Black geographies. *Geography*, *105*(3) pp126–134.

Roberts, M. (2013) *Geography Through Enquiry*. Sheffield: Geographical Association

Rudolph, S., Sriprakash, A., & Gerrard, J. (2018) Knowledge and racial violence: The shine and shadow of 'powerful knowledge'. *Ethics and Education*, *13*(1) pp. 22–38, DOI: 10.10 80/17449642.2018.1428719

Slater, F. (1982) *Learning Through Geography*. London Pearson Education

Sleeter, C. E. (2011) *The Academic and Social Value of Ethnic Studies: A Research Review*. Washington, DC: National Education Association.

Sleeter, C. E., & Zavala, M. (2020) *Transformative Ethnic Studies in Schools: Curriculum, Pedagogy, and Research*. New York: Teachers College Press.

Standish, A. (2021) Geography. In Sehgal-Cuthbert, A. & Standish, A. (2021) *What Should Schools Teach? Disciplines, Subjects and the Pursuit of Truth*. London: UCL Press. www.uclpress.co.uk/products/165025

United States Government Accountability Office. (2015) *Most eighth grade students are not proficient in geography*. www.gao.gov/assets/680/673128.pdf

Walshe, N., & Healy, G. (Eds.). (2021) *Geography Education in the Digital World: Linking Theory and Practice*. Abingdon: Routledge.

Young, M. (Ed.). (1971) *Knowledge and Control: New Directions for the Sociology of Education*. London: Collier Macmillan.

Young M. (2008), *Bringing Knowledge Back in: From Social Constructivism to Social Realism in the Sociology of Education*. Abingdon: Routledge.

Young, M. & Muller, J. (2010) Three Educational Scenarios for the Future: Lessons from the Sociology of Knowledge. *European Journal of Education*, *45*(1) pp11–27

9

YOUNG PEOPLE'S GEOGRAPHIES, SCHOOLING, AND THE CURRICULUM PROBLEM

Where have all the cool places gone?

John Morgan

Introduction

Almost two decades ago, I wrote an opinion piece about the impact that *Cool Places: Geographies of Youth Cultures* (hereafter, *Cool Places*) edited by Tracey Skelton and Gill Valentine had on me as a teacher of geography (Morgan, 2003; Skelton and Valentine, 1998). As part of human geography's cultural turn, *Cool Places* was one of the first geographical books to take seriously young people's cultures and experiences. I can recall the date, place, and time I got hold of my copy. It was 24 October 1997,[1] and I bought it at a bookshop in the north London suburb of Muswell Hill on my way home from a hard day teaching in the sixth-form college where I worked.[2] My nephew had been born earlier that day. The book appeared at just the right time for me. As a teacher, I was feeling intense pressure to 'cover the syllabus' to get students through their exams. The college management was looking for ways to 'improve results' through various performance measures. Though not quite Marx's 'sausage factory', it certainly felt like my students and I were on the educational equivalent of a treadmill.

Over the next few days and weeks, as I eagerly read the chapters in the book, I felt a growing optimism about the possibility of developing an approach to teaching geography that could acknowledge young people's experiences of space and place and recognise their agency and role in shaping and making environments. *Cool Places* built upon the tradition of cultural studies of youth, which argued that young people had 'not been enfranchised by the research conducted on their lives' (p. 21). The book insisted that young people are important actors in their own right but also recognised that their experience of spaces, places, and environment is shaped by the economic and social structures in which they grow up. Here was the socio-spatial dialectic in action – young people making geographies but not in conditions of their own choosing.

DOI: 10.4324/9781003248538-12

Looking back, I can see that my response to *Cool Places* was contingent on, first my training and experience as a geography teacher (1), and second, how the book seemed to define a progressive moment in cultural politics (2). In terms of (1), my reaction to *Cool Places* was about the 'curriculum problem' that I was facing. As a geography teacher, I had been trained – following Norman Graves's influential *Curriculum Planning in Geography* (1979) – to define the curriculum problem as (a) what I was trying to get students to learn; (b) how I was going to manage the operation of making students learn what I had decided was worthwhile; and (c) how I could evaluate what had happened in the learning process so as to improve things next time round. However, over the course of ten years of teaching geography I had refined my definition of the 'problem'. The geography curriculum was ideological. It offered a selective view of society and space and society–nature relations.[3] At the same time, through the 1990s, in part through the impact of postmodernism of both geography and education, there was a loosening of the grip of 'Geography with a capital G' and an increasing focus on 'multiple geographies'. By the time I completed my doctoral study on *Postmodernism and Geography Education* (Morgan, 1998), the modernist definition of the curriculum problem I took from Graves no longer held. For a start, the sense that I was able to exercise authority to command students to learn my definition of geography was a problem – I was positioned as a white, middle-class, straight man, and all that implied. Geographical texts were unstable too, there was no necessary relation between representation and reality, and students were active constructors of geographical knowledge. This too rendered any attempt to evaluate the outcomes of teaching unstable and uncertain. *Cool Places*, occupying a complex position between a 'postmodernism of reaction' and a 'postmodernism of resistance', reflected these developments (McLaren, 1995).

But *Cool Places* was also a product of its time. Susan Bassnett notes that, 'we can see now that 1997 was, in several ways, a watershed year, a year of change, when forces that had been moving below the surface of British society emerged in unexpected ways' (2003: p. 178). Two events stand out. First, the election of a New Labour government in May, bringing to an end 18 years of Conservative rule which had reshaped the economic and social landscape. Second, the accidental death of Diana, Princess of Wales, three months' later, the reaction to which was interpreted by some influential commentators as an indication of how the mood of the nation, its structure of feeling, had changed. *Cool Places* appeared at a moment when Britain's social and cultural geography was being transformed and can productively be read alongside David Morley and Kevin Robins's (2001) edited collection *British Cultural Studies*. That book – which contained essays from geographers including Phil Crang, Peter Jackson, David Sibley, Linda McDowell, and Peter Taylor, as well as Robins himself – reflected changes in the economic, structural, and cultural feeling of a 'young country' (Wilkinson, 1996). It explored how the nations that constituted Britain imagined and represented themselves were changing and explored how a generation of young people – 'Thatcher's children' (Pilcher and Wagg, 1996) – which had grown up with high expectations of property ownership, material possessions, and their individual rights were faring. Taken together, the essays in

British Cultural Studies and *Cool Places* provide a snapshot of how academic commentary was making sense of a society transformed by consumption and in which social relations based on gender, ethnicity, nation, and region were in flux. As a geography teacher, my geographical imagination had been forged slightly earlier, in the context of the bitter conflicts of de-industrialisation and the Miners' Strikes; so *Cool Places* and *British Cultural Studies* seemed to offer new ways forward. As two of the most astute political commentators of the 1990s argued, 'the collapse of political certainty, and our immersion in the impermanence of the contingent, is in one sense liberating' (Coddington and Perryman, 1998: p. 1).[4] This comment, emanating from a loose affiliation of scholars seeking to engage with the question of how Britain can undertake a progressive modernisation to socialism, provides a clue as to my own political project as a geography teacher and educator. Briefly stated, I see how, as the discipline took a spatial turn in the 1950s and 1960s, the most talented human geographers of that generation – here I am thinking of geographers such as Doreen Massey, Kevin Cox, David Harvey, Paul Knox – rejected the new approaches and embraced a form of left-pluralist politics. This eventually led to the discipline's encounter with political and social theory (see Jackson, 1989). School geography, however, largely ignored this political turn. My own interest (Morgan, 2020) is in the possibilities for a school geography informed by what Jones (1983) termed 'curricular socialism'. Curricular socialism refers to the idea that those on the educational left might use the school curriculum to promote socialist values. Huckle (1983) called for a 'socialist geography', but the historic failure of socialism since then means that arguments for a socialist curriculum are quite rare.

Two decades later, *Cool Places* is still on my bookshelf, even if the moment it represented (and which I have just described) has passed. In what follows, I use *Cool Places* as a marker with which to narrate an account of how school geography has handled the question of young people's geographies. I argue that *Cool Places* signalled a crucial moment in how school geography sought to recognise and incorporate young people's geographical experiences into its curriculum thinking. However, if *Cool Places* reflected a postmodern sense of optimism – that social and cultural life were more open and offered more space for agency, especially for young people – then this faded as the 2000s wore on and came to an abrupt end with the global financial crash of 2008. The curriculum problem facing geography teachers in 2022 is very different. The final part of my chapter brings the story up to date.

Culture, young people, and the curriculum

We should not lose sight of the fact that education takes place within a capitalist society (Streeck, 2016). Changes in the form and content of education reflect shifts – however indirectly – in the operations of capitalism. Capitalism faces the problem of how to reproduce itself. Labour is the source of value; so the question is how to reproduce labour. The state takes an interest in this, and as children get older and become young people assumes more responsibility for this. After all, workers do not appear 'ready-made' to take up jobs in offices, shops, and factories. They have to be prepared.[5] Part of the school's job is to make workers, and as the requirements of capital change, so does the curriculum. Consider two examples. In the 1950s and 1960s, the school curriculum reflected assumptions about the 'proper' role for women. However, the 1960s saw an increased demand for female labour, which eventually fed into pressure for equal pay and the passing of the Sex Discrimination Act in 1975. Continued changes in the structure of the economy were reflected in arguments about 'feminization' of the workforce and eventually (and unevenly) reshaped the subjects of the curriculum. Campaigners sought to end divisions between 'boys' subjects and 'girls' subjects and efforts made to encourage girl-friendly approaches in mathematics, science, and geography. A second example is how British capital's solution to labour shortages in the post-war period was to import migrant workers from its former colonies. Assuming that this migration would be temporary, governments – both Labour and Conservative – pursued a policy of assimilation. However, once it became clear that migrants to Britain were 'here to stay', the educational challenge was how to integrate the children of immigrants to life in Britain. This was effected through state-sponsored programmes of multicultural education, culminating in the 1977 Green Paper, which asserted that 'the education appropriate to our Imperial past cannot meet the requirements of modern Britain' (DES, 1977: p. 4).

These examples should be enough to make the point that, for much of the post-war period, education in Britain was a response to a series of challenges in the culture and political economy of the state. A rough sketch of educational change since 1945 would recognise the following elements.

* The expansion of education for all after the 1944 Education Act reflected the demands of working-class people for inclusion following the sacrifices of the Second World War.
* Changes in the structure and organisation of schooling and the curriculum from the mid-1960s were a response to a society experiencing the 'white heat' associated with affluence and modernisation as Britain attempted to transition to a high-skills, knowledge-based economy.
* Consensus about the aims and purposes of schooling crumbled in the face of the oil price shocks of the 1970s and the period of restructuring that followed, leading eventually to a Conservative restoration that veered between modernising the economy and maintaining tradition.
* As the pace of social change quickened in the process of a globalising economy, it looked as though a New Labour government would be able to modernise the structures and institutions of the state but increased levels of inequality and the mounting contradictions of a debt-based model of economic growth. The global financial crash marked the onset of the slow retreat of the neoliberal model.[6]

At each of these stages, young people have been at the sharp end of social and cultural change, and it is the role of schools and teachers to provide them with a 'believable world picture into a curriculum' (Inglis, 1985: p. 22). Of course, any attempt to provide a general account obscures the diversity of children's lives, but it is possible to make some comments about the relationship between capitalism, education, and young people.

The first thing to note is that to be a child or young person in Britain, since 1945, has included being 'schooled'. Children and young people stay in formal education for longer periods and more of each age cohort participate. What's more, higher levels of young people expect to gain educational qualifications. In addition, the general move has been to recognise and acknowledge the experiences and voices of children and young people. This represents a gradual but significant shift in the relations between adults and young people. There is more expectation of democracy, and this reflects, in part, the apparent freedoms associated with a market culture (Hammond and Healy, 2022; Biddulph et al., 2022). Though there remain distinct modes of socialisation for young men and women, there is an assumption of (in theory at least) a variety of roles.

Schools and their curriculum refract these changes in the economic and cultural worlds of children and young people. Though it is uneven, in the post-war period, the general direction has been for schools and teachers to take more account of young people's lives and experiences. This 'educational progressivism' stems from the acknowledgement that children and young people come to classrooms with their own rich 'language in use', which teachers use as the starting point for teaching (Dixon, 1967), and that the boundaries between formal schooling and popular culture are blurred (Hall and Whannel, 1964; Richards, 2010). Moves to recognise and incorporate these experiences into school curricula were part of a larger transformation of cultural authority, with younger teachers more prepared to challenge the sanctity of the 'traditional' curriculum (Holly, 1973). According to Jones (2010: p. 47):

> Thus there developed – *not universally, but in pockets of the state system* – a radical educational culture that questioned the values, traditions, purposes and allegiances of the school and worked on alternative practices, which were to some extent sympathetic to the experiences and meanings of subordinate social groups and therefore committed to the understanding of the school as a place where cultural meanings were brought into relationship with one another and, in the process, remade.

Geography and young people's geographies

It is within this 'long revolution' (Williams, 1961) that the appearance of a collection like *Cool Places* makes sense. Geography, as a discipline, had its roots in imperialist expansion. In Britain, it emerged as one of the 'English' subjects, developed to respond to a wider economic and political crisis as Britain faced international competition (Doyle, 1989). It was patrician, middle class, and male dominated, aligned with a popular culture that celebrated a masculinity based on exploration, adventure, and the stiff upper lip (Phillips, 1996).[7] Walford's (2000) *Geography in British Schools 1850–2000* provides a sense of the social milieu of the upper middle-class men who shaped the subject in schools and their attitudes to landscapes and places.

This meant that geography – both academic and school versions – was slow to acknowledge the geographical lives and experiences of young people. Despite important changes in the post-Second World War period, this continued. The 'revolutions' taking place at the frontiers of the academic discipline in the 1960s and 1970s coincided with the expansion of education and the reorganisation of secondary schooling. However, whereas in other parts of the curriculum the response to cultural change was to acknowledge and find ways to work with students' experiences, school geography took another tack. The quantitative and statistical revolutions were translated into a 'new geography' that could be taught in schools. School geography excluded nature from its models of locational analysis, marginalised politics in the rush to represent society as a well-oiled machine, and ignored how people experienced the reality of social and economic change (Huckle, 1983).

All this meant that, when geography teachers in Britain were faced with comprehensive reorganisation and the Raising of the School Leaving Age (ROSLA) to 16 in 1972, for many it came as a shock. 'Relevance' became a code word for 'working-class' students. As the editorial for the first issue of *Teaching Geography* (a journal aimed at teachers of geography published by the Geographical Association) admitted that 'we still know remarkably little about the ways in which children and young people learn geography, or indeed anything else' (Bailey, 1975: p. 1).

This is not the full story, however, and through the 1970s and into the 1980s, there were moves to recognise the geographical experiences of children and young people as part of what Thomson (2013) calls the 'radicalization of the post-war landscape settlement'. In the 1960s and 1970s, the physical landscape of towns was being reshaped through redevelopment schemes, new shopping centres, ring roads, and traffic schemes. These changes were the stuff of geography lessons and prompted a response. The social environment was changing too, with the 'rediscovery of poverty' from the mid-1960s and a growing awareness of environmental impacts on pupils' educational progress.[8] These fed into the 'geography of social concern' associated with welfare geography (Knox, 1975).

These issues were addressed by a newly formed 'Environmental education' (Martin and Wheeler, 1975), which went beyond the traditional concern with unique and special (invariably rural) sites to focus on the everyday places and spaces in which children lived.[9] The student-centred approaches to teaching and learning it favoured acknowledged that young people were developing as political citizens. The Skeffington Report (1969) – *People and Planning* – called for participation in planning. It stated that education about town planning should be 'part of the way in which all secondary schools make children conscious of their future civic duties' (para. 245). This extended to teacher training, which should include studies of 'the philosophy of town and country planning' (para. 247).

In response to the Skeffington Report, the Town and Country Planning Association appointed Colin Ward and Anthony Fyson (a geography teacher) as its environmental education officers. Fyson proposed the term 'Streetwork', which had a rather different focus than typical fieldwork (Ward and Fyson, 1973). In a book – *Changing the Street* – written for pupils, Fyson (1976) explained:

> This book is about streets and the people who live in them. It is not about important streets, not about famous buildings; it is about the kind of streets most of us live in, the streets we use all our lives, and what happens to them makes a difference.
>
> *(p. 1)*

Here we have a call for an educational practice that makes links between the class-rooms, communities, and 'everyday life'. These 'grassroots' initiatives – which, as Fyson's quote suggests, were 'activist' in nature – were able to flourish in the context of 'educational Keynesianism' with its state-sponsored curriculum development which spawned a series of educational innovations, including the Humanities Curriculum Project, the integrated Humanities projects, and the Geography for the Young School Leaver (GYSL). All these initiatives favoured a teaching approach in which students were encouraged to examine 'evidence', study relevant 'issues', and acknowledge the role of 'values'. As such, they fell under the umbrella of 'progressive education' and were often directly critical of the 'new geography', which, it was argued, consisted of 'mind-bending exercises in statistics largely irrelevant to the urgent problems confronting the environment we live in' (Wheeler, 1975, p. 11). According to this account, geography – the one subject in the curriculum that the public might expect to teach children about the importance of place – had, for the last two decades, failed to do so (Goodey, 1982).

Geography educators' reception of 'streetwork' and issues-based approaches were mixed. Walford (2000) offered faint praise for Ward and Fyson's *Bulletin of Environmental Education*, describing it as 'quaint', and the editors of a collection that included Michael Storm's (1971) article on the need for a relevant and engaged approach to local issues asked readers to reflect on the loss of rationality and involvement as privileging emotion over thought. Others were more welcoming, such as David Wright's (1983) letter to *Teaching Geography*, which called for more time studying urban pavements and less focus on limestone pavements!

Whilst there was an element of patch-protection in these geography educators' responses, marked by a desire to resist subject integration (Williams, 1976), I think it can also be interpreted as a suspicion of a political tradition that drew variously upon strands of social ameliorism, anarchism, and Fabianism to challenge the assumptions of post-war social democracy (Worpole, 1999). In their most developed forms, Streetwork and Environmental Education represented a rejection of the geographer's faith in a consensus view of society, the ability to manage conflict, and to apply rational planning decision-making to order the economic and social landscape.[10] The argument was that an academic and sanitised view of the planning process was increasingly out of kilter with the lived landscapes of childhood and young people as the economic boom lost momentum and the post-war settlement began to lose coherence.

Despite these reservations, the broad progressive educational climate of the time ensured that calls to teach a more 'human' geography were acknowledged. 'Streetwork' and 'Environmental Education' resonated with the growing interest in behavioural geography, which, taking its cue from J.K. Wright's (1947) observation that 'geography has to do in large degree with subjective concepts', came to play a significant part in curriculum development in geography education.

Behavioural geography proved attractive for those geography educators who were taking an interest in cognitive psychology and its application to curriculum planning. It introduced concepts such as 'perception', 'cognition', and 'beliefs' and was given a further boost by the development of new survey techniques (such as 'Likert scales' and 'Repertory Grids'). These offered ways to measure and quantify attitudes and values and linked to the notion of constructionism through which individuals accommodate and assimilate new information from their environment, thereby shaping their values (Fien and Slater, 1983; Slater, 1982) – sets of ideas that shaped how they think about the world and act towards it (Watson, 1989; Watson, 1977). Whilst much behavioural geography could be incorporated within the 'positivist' approach to geography teaching, it also raised the question of what represented a 'valued environment'. This might, as Gold and Burgess (1982) suggest, include not only 'those most cherished buildings and landscapes that are part of the national heritage and which enjoy widespread protection' but also other environments 'that supply the backcloth for everyday life' (p. 1).

Thus, behavioural geography might explore the ways in which ordinary people made sense of and related to environments that were considered by experts as lacking value or popular landscapes, such as football stadia, cinemas, or allotments, frequently outside of the interest of geographers. In this way, mainstream school geography found ways to acknowledge and come to terms with the lived experiences and 'geographies of young people'.

These initiatives did not last though. Financial pressures meant local initiatives such as Streetwork and the Bulletin of Environmental Education were wound up. Behavioural geography lost ground as academic geographers favoured 'new models' based on political economy and social and cultural theory. These changes were moreover linked to the changing educational politics of the 1980s, as teachers lost control of the classroom (Lowe, 2007).

Neoliberal capitalism and the turn to culture

In the 1980s, in response to the crisis of social democratic education, policy and practice took a sharp 'right turn' (Jones, 1989). Robins and Webster (1989) provide a succinct account of this shift. Faced with an accumulation crisis, British capitalism sought to recalibrate its demand for labour. Schools and universities – and by extension teachers – were charged with (and found guilty of) perpetuating an anti-industrial spirit (Weiner, 1981). The shift from Fordism to post-Fordism would require more flexible, agile responses from industry and labour, and this would mean that the aims and purposes of the curriculum required reform (Helsby, 1999).

The ruling Conservative Party was, however, divided on education policy between 'cultural restorationists' and 'curriculum modernizers' (Ball, 1994), which itself reflected the wider division in British culture in the 1990s between 'heritage' and 'enterprise' (Corner and Harvey, 1991). Children and young people were at the centre of a series of politically motivated 'moral panics' around the widespread fear of youth unemployment, the pernicious effects of media culture, and sexuality. For Conservatives, this requited that the boundaries of the nation were to be policed more firmly, notably through a 'back to basics' National Curriculum (Quicke, 1989).

This is a familiar tale, but it is important to remember that at every stage, conservative hegemony was contested by educators and teachers committed to other versions of the 'schooled child'. Indeed, in the long decade of Conservative government (1979–1997), the boundaries of children's rights and children's voices were extended. For example, the UK signed the UN Convention on the Rights of the Child (UNCRC, 1989), and the UK Agenda for Children (CRDU, 1994) sought to assess the extent to which governments complied with its principles.

Over time, the traditionalist arguments about the curriculum lost ground. In part, this was to do with longer-term shifts in the culture – a form of de-traditonalisation, which meant that by the mid-1990s, conservativism found itself without a home (McRobbie, 1992). During the 1990s and 2000s, there emerged a cultural and educational formation that focused on children's agency, student voice, ideas of participation and citizenship, all underpinned by the idea that knowledge is a co-construction. This educational formation created enough space for geography educators to draw upon perspectives and insights from the growing interest in children's geographies and young people's geographies (McKendrick, 2022). This radical educational space was the result of two forces. First, in the 1980s, human geography took a distinctly left turn (Thrift, 1992). The new models in geography were those that drew from neo-Marxist political economy and social and eventually cultural theory. Academia was a culture in contraflow to the new right (Anderson, 1990). One important aspect of this was a focus on children and young people's geographies, which worked with the idea that childhood is a social construction rather than a 'natural' or biological condition (James et al., 1998). This resonated with researchers in education and geography education and especially those involved in primary geography, where the progressive notion of the child as active agent was more established (Catling and Pike, 2022). Second, the cultural mood in schools had changed to the extent that on the eve of the election of a new Labour government, Jones (1996) could assert that:

> The school, which was for so long thought of as the site of a simple transmission of a society's roles and values has now come to be perceived as a place where students learn how to learn – how to develop the capacity to generate new meanings. Such a redefinition alters the status of school knowledge,

which loses much of the unquestioned character of its authority. More than that, it provides those who work in education with more complex and demanding roles.

(p. 6)

The election of a new Labour government in 1997 inaugurated a period of educational experimentation and innovation that was led by a variety of 'think-tanks', 'knowledge labs', corporations, and 'third sector' organisations, many of which were linked with educators working in universities. Examples included the RSA's *Opening Minds*, Futurelab's *Enquiring Minds* and the Paul Hamblyn Foundation's *Learning Futures*. What linked these diverse projects is that they all were based on a critique of traditional models of schooling, they all saw teachers less as transmitters of knowledge and more as facilitators of student learning; they all worked with the idea that children's lives were changing, and that there was more room for agency.[11] In short, they drew upon the idea that there were new cultures of learning or a 'new progressivism' (Moore, 2000).

This brings us to the moment – 1997 – at which *Cool Places* appeared. In terms of geography as an academic subject, it reflected the way in which geography had become an important part of the British intellectual left's attempts to come to terms with the changes that had taken place in political economy since the 1980s. In terms of school geography, it highlighted how forms of curriculum and pedagogy had emerged that sought to acknowledge young people's geographical experiences and find ways to recognise and make use of them in the classroom. In the concluding section of this chapter, I will focus on the 'curriculum problem' teachers of geography face today.

The curriculum problem today

Cool Places represented an important moment in the rethinking of the youth question (Cohen, 1996). It reflected the long revolution that had worked its way through education and democracy (Stevenson, 2003) and, in acknowledging the importance of young people's voice and participation, drew upon important traditions in political education (Tapper and Salter, 1978). The most direct impact on school geography of the approach advanced in *Cool Places* was the Young People's Geographies project.

Things have moved on though. If *Cool Places* appeared at a time of relative optimism about the prospects for what Michael Rustin (2019) terms 'progressive agency', this is hardly the case today. If we were to search for titles that encapsulate the predicament of contemporary youth, we might suggest that the paradigmatic texts for young people growing up today could be *Non-Stop Inertia* (Southwood, 2010), *Planet on Fire* (Lawrence and Laybourn-Langton, 2021), or even *The Worst is Yet to Come* (Fleming, 2019). Pickard offers a sober assessment:

Today, a majority of young people in Britain have fewer opportunities and endure a poorer quality of life than their parents and grandparents. The traditional markers of entry into adulthood have been largely dissolved and seem unobtainable to a substantial proportion of young people in contemporary advanced democracies. Rather than climbing the social ladder, numerous young people are sliding down it into precarity, insecurity and debt, i.e. undergoing a 'downgrading' of their social situation and downward social mobility compared to the upward mobility of older generations.

(Pickard, 2019: p. 7)

What then are the possibilities for a geography education that plays its part in an education of self-realisation and self-discovery, a dimension of lived experience? In the last decade, education policy has sought to establish 'proper' culture (based on what many argue is an elite culture) as the basis for the school curriculum. This is particularly evident in calls for Ofsted[12] to assess how far schools provide students with access to 'cultural capital', a term which, far from its origins in the work of the French sociologist Pierre Bourdieu, is simply rendered as respected and valued knowledge. This move has been widely criticised because such core knowledge is based on a canon of great literature and art reminiscent of the Victorian school inspector Matthew Arnold's the 'best that has been thought and said'.

Although individual teachers and school geography departments may be working out how to connect with young people's experiences, at the present time there is little 'official' support for such approaches. However, as I have attempted to show in this chapter, there exists an important strand of work in school geography in Britain which seeks out and engages with the geographical experiences of young people. The challenge is to find ways to advance this work in the current 'mean times', when the prospects for progressive agency seem bleak. That is why *Cool Places* still matters.

Notes

1 Lest this cause confusion: academic books often appear on the bookshelf slightly earlier than their printed publication dates.
2 In England and Wales, a 'sixth-form' college caters for students aged 16–19.
3 See Apple (1979), Gilbert (1984), Huckle (1985).
4 Coddington and Perryman were important figures in the group Signs of the Times, which provided a forum for the progressive left in the 1990s. It sought to develop the new times analysis associated with Marxism Today and deepen its insights as the prospect of New Labour gaining power increased. Signs of the Times sought to develop 'an understanding of the profound changes which are redrawing the political and cultural map'. These are explored in three edited collections (see Perryman, 1992, Perryman, 1996, and Coddington and Perryman, 1998). I attended meetings and conferences in the mid-1990s, as I was writing my doctoral thesis.
5 Hence Paul Willis's (1977) *Learning to Labour* and Katharyne Mitchell's (2018) *Making Workers*.
6 The best account of these changes, and one that seeks to provide a 'four nations' approach to education in Britain, is Jones (2015).
7 It is important to remember that, in Britain, geography as a university subject was established to supply teachers to schools, a situation which pertained until the early 1950s.

8 The 'rediscovery of poverty' refers to the arguments that surrounded the publication of Brain Abel-Smith and Peter Townshend's *The Poor and the Poorest* in December 1965. This was the moment when it became clear, in social policy and administration, that, despite post-war affluence, pockets of poverty remained. The Child Poverty Action Group (CPAG) was formed in the same year.

9 I am using the term 'Environmental Education' here to refer to a very specific educational formation, one that stresses the importance of seeing environmentalism as 'ordinary'. It is about the experience of being in places, the historic buildings, the feel of the place, the architecture – what might be called a 'sense of place'. It grew out a coming together of the changes taking place in the urban landscape in the post-war period and a recognition that schools and classrooms where children and young people can make sense of that experience.

10 Recall David Harvey's (1974: p. 22) acerbic comment:

> The self-image of the geographer at work appears to be one of doing good. Tune into any discussion among geographers and as likely as not the discussion unfolds from the standpoint of the benevolent bureaucrat, a person who knows better than other people and will therefore make better decisions for others than they will be able to make for themselves.

11 I had some personal experience of this educational formation. Between 2005 and 2008, I worked at Futurelab – a not-for-profit educational organisation that had as its goal the transformation of learning through the use of technology and creative pedagogies. After working as a teacher educator in a university, I suddenly found myself sitting in meetings where policymakers exhorted the benefits of learner-centred teaching, personalisation, and student voice. They saw the content-led curriculum based around subjects as outmoded and reactionary.

12 Ofsted (or the Office for Standards in Education) is the body that inspects schools in England.

References

Anderson, P. (1990) A culture in contraflow – 2. *New Left Review*, 182, pp85–137.

Bailey, P. (1975) 'Editorial'. *Teaching Geography*, 1(1) pp4–5.

Ball, S. (1994) *Education reform: a post-structural approach*. London: Routledge.

Bassnett, S. (2003) 'Afterword: Studying British cultures, 2003'. In. S. Bassnett (ed.) *Studying British cultures*. 2nd edition. London: Routledge.

Biddulph, M. Hopkins, P. Tate, S. (2022) 'Connecting Children's and Young People's Geographies and Geography Education: why this matters to and for children, education, and society'. In L. Hammond, M. Biddulph, S. Catling, J.H. McKendrick (eds) *Children, education and geography: rethinking intersections*. Abingdon: Routledge.

Catling, S. Pike, S. (2022) 'Becoming acquainted: exploring young(er) children's geographies'. In L. Hammond, M. Biddulph, S. Catling, J.H. McKendrick (eds) *Children, education and geography: rethinking intersections*. Abingdon: Routledge.

Children's Rights Development Unit. (1994) *UK agenda for children: A systematic analysis of the extent to which law, policy and practice in the UK complies with the principles and standards contained in the UN convention on the rights of the child*. London: CRDU.

Coddington, A. Perryman, M. (eds.). (1998) *The moderniser's dilemma: radical politics in the age of Blair*. London: Lawrence and Wishart.

Cohen, P. (1996) *Rethinking the youth question*. Basingstoke: Macmillan.

Corner, J. Harvey, S. (eds.). (1991) *Enterprise and heritage: crosscurrents of national culture*. London: Routledge.

Department of Education and Science. (1977) *Education in schools*. London: HMSO.

Dixon, J. (1967) *Growth through English*. London: NATE.

Doyle, B. (1989) *English and Englishness*. London: Routledge.

Fien, J. Slater, F. (1983) Behavioural geography. In Huckle J. (ed.) *Geographical education: reflection and action*. Oxford: Oxford University Press.

Fleming, P. (2019) *The worst is yet to come: a post-capitalist survival guide*. London: Repeater Books.

Fyson, A. (1976) *Change the street*. Oxford: Oxford University Press.

Gold, J. Burgess, J. (eds.). (1982) *Valued environments*. London: George Allen and Unwin.

Goodey, B. (1982) Values in place: interpretations and implications from Bedford. In J. Gold, J. Burgess (eds) *Valued environments*. London: George Allen and Unwin.

Graves, N. (1979) *Curriculum planning in geography*. London: Heinemann Educational Books.

Hall, S. Whannell, P. (1964) *The popular arts*. London: Hutchinson Educational.

Hammond, L. Healy, G. (2022) 'Student voice, democratic education and geography: reflecting on the findings of a survey of undergraduate geography students'. In L. Hammond, M. Biddulph, S. Catling, J.H. McKendrick (eds) *Children, education and geography: rethinking intersections*. Abingdon: Routledge.

Harvey, D. (1974) What kind of geography for what kind of public policy? *Transactions of the Institute of British Geographers*, 63, pp. 18–34.

Helsby, G. (1999) *Changing teachers' work: the 'reform' of secondary schooling*. Buckingham: Open University Press.

Holly, D. (1973) *Beyond curriculum*. London: Hart-Davis MacGibbon.

Huckle, J. (1983) *Geographical Education: reflection and action*. Oxford: Oxford University Press.

Huckle, J. (1985) Geography and schooling. In R. Johnston (ed.). *The future of geography*. London: Methuen.

Inglis, F. (1985) *The management of ignorance: a political theory of the curriculum*. Oxford: Basil Blackwell.

Jackson, P. (1989) *Maps of meaning*. London: Unwin and Hyman.

James, A. Jenks, C. Prout, A. (1998) *Theorizing childhood*. Oxford: Wiley.

Jones, K. (1983) *Beyond progressive education*. Basingstoke: Macmillan.

Jones, K. (1989) *Right turn: the Conservative revolution in education*. London: Hutchinson Radius.

Jones, K. (1996) Cultural politics and education in the 1990s. In R. Hatcher. K. Jones (eds) *Education after the conservatives: the response ot the new agenda of reform*. Stoke-on-Trent: Trentham Books.

Jones, K. (2010) Schooling and culture. In M. Higgins, C. Smith, J. Storey (eds) *Cambridge companion to modern British culture*. Cambridge: Cambridge University Press.

Jones, K. (2015) *Education in Britain: 1944 to the present*. 2nd edition. Cambridge: Polity.

Knox, P. (1975) *Social well-being: a spatial perspective*. Oxford: Oxford University Press.

Lawrence, M. Laybourn-Langton, L. (2021) *Planet on Fire: a manifesto for the age of environmental breakdown*. London: Verso.

Lowe, R. (2007) *The death of progressive education: how teachers lost control of the classroom*. London: Routledge.

Martin, G. Wheeler, K. (eds.). (1975) *Insights into environmental education*. Edinburgh: Oliver and Boyd.

McRobbie, A. (1992) Revenge of the 60s. *Marxism Today*, January.

McLaren, P. (1995) *Critical pedagogy and predatory culture*. London: Routledge.

McKendrick, J. H. (2022) Children's geographies and schools: beyond the mandated curriculum. In L. Hammond, M. Biddulph, S. Catling, J.H. McKendrick (eds) *Children, education and geography: rethinking intersections*. Abingdon: Routledge.

Mitchell, K. (2018) *Making workers: radical geographies of education*. London: Pluto Press.

Moore, A. (2000) *Teaching and learning: Pedagogy, curriculum and culture*. London: Routledge.

Morgan, J. (1998) *The implications of postmodernism for school geography: a discussion*. Unpublished doctoral dissertation, University of London.

Morgan, J. (2003) On Cool places by Tracey Skelton and Gill Valentine. In J. McKendrick (ed.) *First steps: a primer on the geographies of children and youth*. London: Royal Geographical Society.

Morgan, J. (2020) The prospects for knowledge socialism in one country. In M. Adrian Peters, T. Besley, P. Jandrić, X. Zhu (eds.), *Knowledge socialism: the rise of peer production: Collegiality, collaboration and peer production*. Singapore: Springer.

Morley, D. Robins, K. (eds.). (2001) *British cultural studies*. Oxford: Oxford University Press.

Perryman, M. (ed.) (1992) *Altered states: postmodernism, politics, culture*. London: Lawrence and Wishart.

Perryman, M. (ed.) (1996) *The Blair Agenda*. London: Lawrence and Wishart.

Phillips, R. (1996) *Mapping men and empire*. London: Routledge.

Pickard, S. (2019) *Politics, protest and young people: Political participation and dissent in 21st century Britain*. Basingstoke: Palgrave Macmillan.

Pilcher, J. Wagg, S. (eds.) (1996) *Thatcher's children? Politics, childhood and society in the 1980s and 1990s*. Abingdon: Routledge Falmer.

Quicke, J. (1989) 'The new right and education'. In. B. Moon, P. Murphy. J. Raynor (eds) *Policies for the curriculum*. London: Hodder and Stoughton.

Richards, C. (2010) *Young people, popular culture and education*. London: Continuum.

Robins, K. Webster, F. (1989) *The technical fix: education, computers and industry*. Basingstoke: Macmillan.

Rustin, M. (2019) The question of progressive agency: What kinds of agency are most likely to bring about the changes in society we so urgently need? *Soundings: A Journal of Politics and Culture, 72*, pp. 48–64.

Skelton, T. Valentine, G. (eds.) (1998) *Cool Places: Geographies of Youth Cultures*. London: Routledge.

Slater, F. (1982) *Learning through geography*. London: Heinemann.

Southwood, I. (2010) *Non-stop inertia*. London: Zero Books.

Stevenson, N. (2003) *Cultural citizenship: cosmopolitan questions*. Buckingham: Open University Press.

Storm, M. (1971) Schools and the community: an issues-based approach. *Bulletin of Environmental Education*, 1.

Streeck, W. (2016) *How will capitalism end?* London: Verso.

The Skeffington Report (1969) *People and planning: report of the committee on public participation in planning*. London: HMSO.

Tapper, T. and Salter, B. (1978) *Education and the political order: changing patterns of class control*. Basingstoke: Macmillan.

Thomson, M. (2013) *Lost freedom: the landscape of the child and the British post-war settlement*. Oxford: Oxford University Press.

Thrift, N. (1992) Light out of darkness? Critical social theory in 1980s Britain. In: P. Cloke (ed.) *Policy and change in Thatcher's Britain*. Oxford: Pergamon Press.

UNCRC. (1989) *Convention on the rights of the child*. Geneva: United Nations Human Rights.

Walford, R. (2000) *Geography in British schools, 1850–2000: making a world of difference*. London: Woburn.

Ward, C. Fyson, A. (1973) *Streetwork: the exploding school*. London: Routledge Kegan and Paul.

Watson, J.W. (1977) On the teaching of value geography. *Geography* 62(3) pp198–204

Watson, W. (1989) 'People, prejudice and place'. In. F. Boal. D. Livingstone (eds) *The behavioural environment: essays in reflection, application and re-evaluation.* London: Routledge.

Weiner, M. (1981) *English culture and the decline of the industrial spirit, 1850–1980.* Cambridge: Cambridge University Press.

Wheeler, K. (1975) The genesis of environmental education. In: G. Martin. K. Wheeler (eds) *Insights into environmental education.* Edinburgh: Oliver and Boyd.

Wilkinson, H. (1996) 'The making of a young country'. In M. Perryman (ed.) *The Blair agenda.* London: Lawrence and Wishart.

Williams, M. (ed.) (1976) *Geography and the integrated curriculum: a reader.* London: Heinemann Educational Books.

Williams, R. (1961) *The Long Revolution.* Harmondsworth: Penguin.

Willis, P. (1977) *Learning to Labour: how working-class kids get working-class jobs.* London: Routledge and Kegan Paul.

Worpole, K. (ed.) (1999) *Richer futures: fashioning a new politics.* London: Earthscan.

Wright, D. (1983) 'Viewpoint: The road to Malham Tarn'. *Teaching Geography*, 8(3), pp. 139–41.

Wright, J.K. (1947) *Terrae incognitae*: the place of the imagination in geography. *Annals of the Association of American Geographers*, 37(1) pp1–15.

SECTION III
Progressive geographies in education

10

DE/COLONISING THE (GEOGRAPHY) CURRICULUM

Fatima Pirbhai-Illich and Fran Martin

Introduction

In this chapter we explore the potential for de/colonising educational relations in the context of geography education in two countries: Canada (within the context of social studies) and the UK (where it is taught as a discrete subject). We have previously shown how the discipline of geography cannot be understood without paying attention to its relation to, and emergence from, colonialism (Martin and Pirbhai-Illich, 2017). An understanding of this is needed before it is possible to consider how it might be decolonised. Calls to systematically decolonise and indigenise higher education in Canada have been ongoing following the findings of the Truth and Reconciliation Commission (2015), and in the UK since the #RhodesMustFall and #FeesMustFall movement led by student activists in South Africa (Kumalo, 2021). In the UK, the decolonisation of school education gained traction only in 2020, following the murder of George Floyd (Okolosie, 2020). Whilst the desire to decolonise the school curriculum is understandable, to do so without having previously engaged with the extensive scholarship on decolonisation risks such actions becoming reductionist and performative, repeating the very harms that they are intended to address.

In what follows we therefore begin by situating ourselves in relation with colonialism, by providing some clarity over the distinction between colonialism and coloniality and by identifying some of the ways in which coloniality and colonial discourses have structured the world, including the world of geography education. We argue that a critical analysis of the coloniality of geography and education is an essential prerequisite to considering how geography education might be decolonised. Furthermore, we distinguish between decolonisation and de/colonisation and explain why we prefer the latter term with the forward slash and explore how a de/colonial approach might work in practice, referring to an example from each

DOI: 10.4324/9781003248538-14

of our contexts. We conclude by arguing that de/colonising educational *relations*, based on a critical understanding of how a non-hierarchical, non-oppositional, interconnected, interrelational, and interdependent understanding of the world, is a necessary foundation from which it might be possible to de/colonise the curriculum.

Positionalities

Inherent in a critical, intercultural, and interrelational approach is the need to understand one's identity and the subjectivities and positionalities that come with it. We therefore begin by sharing aspects of our identities and the sociocultural, historical, and geopolitical contexts in which they have been formed.

Being a bystander

I am Fatmakhanu (fatima) Shamshudin Pirbhai Sunderji Samji Ladha Kurji Thoba Jessani-Illich. My ancestors originate from Kutch, India, and migrated to Zanzibar and Tanganyika, East Africa, where they lived as members of the Indian diaspora. In 1961, Tanganyika became independent and, in 1964, Tanganyika and Zanzibar united to form the United Republic of Tanzania. At this time, I learnt that we were not welcome there even though it had been our family home for several generations. During the civil unrest, my mother and I immigrated to Canada; however, my grandfather and father stayed behind and so the land and familial ties to Tanzania remained, albeit as an outsider looking in. In Canada, I quickly learnt that my brown body and cultural ways of being and knowing were not welcome there either. Much later, I also learnt that Canada was in fact Turtle Island and that the peoples of First Nations descent were the rightful custodians of this land. For more than four decades, I have been a citizen of a country where I have been made to feel that I do not belong. The centrality of relations to land that is part of both Indigenous and white-settler identities is not part of my identity. I have a sense of being a bystander to this way of thinking because of the transient nature of diasporic relations to land. 'Home' as a relation to land is always temporary to me because 'a return to "Home" is an eternal impossibility, [while] a reframing of home is a continuous negotiation' (Bhattacharya, 2018: p. 15). My 'home' is a diasporic space of in-betweenness, a space without land.

Rooted in the landscape

I am Frances (Fran) Elizabeth Martin. I was born on the family farm in the southeast of England, and lived there all my childhood. As children, we had the freedom to roam in the fields and woods and never had to question our right to be there or our right of access to the land. It was a carefree existence without external conflict; a privileged existence of entitlement, permanence, and authority. As a white, middle-class Briton, daughter of a farmer and a landowner, I have not had to consider the politics of the countryside. I have always felt I belong to the land and that the land, in a sense, belongs to me. It is only in the last 10–15 years that I have become consciously aware that the British countryside is not available to all, that it is subject to a racialised discourse of who 'belongs' there and who does not: 'For many black and Asian Britons, rural Britain is, literally, another country' (Prasad, 2004: n.p.). In contrast, my relation to land is one of permanence and an integral part of my identity. I have a strong place attachment, a positive affective bond with the English landscape, that enhances my sense of individual well-being but also locates it within the wider social world, giving me a strong, stable sense of community and citizenship (Qazimi, 2014).

Our identities and colonialism

Our narratives demonstrate how our histories and geographies have been pro-foundly influenced by colonialism. These differences can be seen in our connec-tions to land, our status as citizens, and our cultural ways of being, doing, and knowing – all of which contribute to our identities and our differing feelings of belonging, sense of permanence, and acceptance in the places and spaces we inhabit. In educational contexts, where multiple identities come together and interact, their histories and geographies will influence how everyone relates to each other and to the curriculum. It is for this reason that we believe any project in de/colonisation must start with developing an awareness and understanding of the nature of coloniality and its influence on identities and thus educational relationships.

Colonialism, coloniality, and colonial discourses

In this section we distinguish between colonialism and coloniality and identify some of the colonial discourses that are embedded in Euro-Western modernity which have influenced the structure of geography as a discipline and the ways in which it constructs knowledge about the world.

Colonialism is the ideology of superiority that led to the Western imperial/colonial expansion from Europe across the world with the intention of acquiring full or partial political control over other countries, occupying it with settlers, and exploiting it economically. Colonising nations violently conquered other nations, subjugating the colonised population, who were forced to erase their own ways of being, doing, and valuing and to adopt the language and cultural values of the colonisers.

Coloniality is the *underlying logic* of all Euro-Western modern/colonial imperial-isms (Quijano and Ennis, 2000) that is the ongoing legacy of colonialism. It is a knowledge system that classifies phenomena on the basis of 'objective' character-istics,[1] placing them into categories that are arranged in a hierarchical structure. Coloniality is perpetuated through institutions of power (e.g., legal systems, educa-tion) which privilege Euro-Western ways of being, doing, and knowing on a global scale (Grosfoguel, 2011). Grosfoguel identifies 15 'entangled, global hierarchies', including:

- a racial/ethnic hierarchy that privileges European people over non-European people;
- a gender hierarchy that privileges males over females;
- a sexual hierarchy that privileges heterosexuals over LGBTQ+;
- a spiritual hierarchy that privileges Christians over non-Christian/non-Western spiritualities;
- an epistemic hierarchy that privileges Western knowledge over non-Western knowledges – as institutionalised in education systems; and
- a pedagogical hierarchy where Cartesian Western forms of pedagogy are considered superior over non-Western forms and practices of pedagogy.

Colonial discourses are based on a binary logic of classification and separation that is oppositional. Colonial discourses produce white, Euro-Western peoples and cultures as rational, modern, advanced, and civilised in opposition to non-white, non-Euro-Western peoples and cultures who are categorised as magical, exotic, violent, backward, and uncivilised. This logic of separation creates an either/or mentality that denies the possibility of both/and which, as we will demonstrate later, leads to a replacement mentality that is fearful of the inclusion of difference.

Coloniality of power and the white possessive

Colonial power relations are both discursive and geographical (Quijano and Ennis, 2000). They rest not only on the 'texts, systems of signification, and procedures of knowledge generation' described earlier but also on 'colonialism's basic geographical dispossessions of the colonised' (Harris, 2004: p. 1).

Colonisation rested on the idea of taking possession – of land, resources, and bodies – the corollary of which is dispossession of the land, resources, and bodies of those who are colonised. Ladson-Billings (2009) argues that 'The significance of property ownership as a prerequisite to citizenship was tied to the British notion that only people who owned the country, not merely those who lived in it, were eligible to make decisions about it' (p. 25). Moreton-Robinson (2015) describes 'white possession' as a logic that extends to possession of identities (the right to define who is white), bodies (the right to claim who can be a free citizen), and to institutions of power that legitimate and reproduce those rights through legal and education systems. The logic of [dis]possession extends the original violence of colonialism through the creation of systems of ownership, asset, and privilege that are afforded to those who are classified as white, Western, European, male, and all the other dimensions of identity that Grosfoguel (2011) are listed in the global, entangled hierarchies we referred to earlier. Coloniality is therefore inextricably tied to land as well as social relations. Territories have been possessed and renamed to reflect the language and culture of the colonisers and a vocabulary of belonging – insider/outsider, included/excluded, citizen/immigrant developed. This same logic, through its construction of spaces as white and for exclusive use, creates a fear of any encroachment by another because, when boundaries are seen as fixed, stable, and impermeable, any attempt to move into those spaces by the Other is likely to be understood as a move to takeover, a replacement of one with another.

The coloniality of geography

One of the biggest challenges teachers of geography have to contend with is that the history of geography and that of colonial empires is inextricably entangled. In 1998, Cheryl McEwan stated that 'geography has not been transformed into a postcolonial discipline. . . . [it] remains overwhelmingly white, Eurocentric and, in some cases, colonial in terms of its practitioners and the subject of its inquiry' (McEwan, 1998: p. 1). Twenty years later, although there have been efforts to develop 'postcolonial geographies' (Jazeel, 2019), Marcin Stanek argues that, 'The ongoing coloniality of geographical knowledge production is not only widely accepted, it is also now every geographer's problem' and asks, 'To what extent are we, as students, teachers, researchers, lecturers, and professors, relating to and working with each other every day in a way that is underpinned by lessons learned from colonial oppressions and grassroots decolonial struggles?' (Stanek, 2019: p. 9–10).

Geography is 'an instrument of domination in the service of conquest. The act of naming a territory and putting it on a map signifies appropriating it' (de Rugy, 2020: n.p.). Coloniality is materially embedded in the land when spaces and places are planned. Geographies of separation took place in the partition of Africa, and the formation of India, Pakistan, and Bangladesh.

In settler colonial nations such as Canada, places were divided along racialised lines through the British system of Indian reservations (Hanson, 2020). The construction of cities and towns also constituted a core activity in the building of the settler-colonial nation, acting as 'a primary mechanism operationalising the spatial and economic dispossession of colonised peoples' and rendering 'the urban as a place not Indigenous, profoundly spatially and temporally disconnected from Indigenous histories and geographies' (Porter and Yiftachel, 2019: p. 177).

Places are thus not neutral; how they are constructed and the boundaries that are drawn – whether real or imaginary – are forms of apartheid creating places of exclusion on the basis of 'race',[2] class, ethnicity, sexuality, religion, and [dis] ability, resulting in spatial hierarchies that reinforce and reproduce sociocultural hierarchies.

The coloniality of geography: the structure of the discipline

As we mentioned earlier, a core aspect of coloniality is the hierarchical classification of phenomena into groups, such as the classification of knowledge into disciplinary silos where science, technology, engineering, and maths (STEM) are valued more highly than the humanities or the arts. Within the English school system,

The coloniality of Geography: categorization and fragmentation

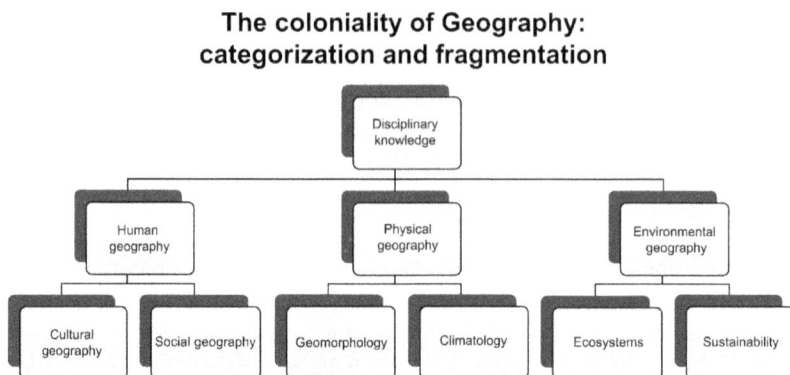

FIGURE 10.1 An indicative example of the fragmented nature of the discipline of geography

this hierarchy is evident in the allocation of greater time and resources to STEM subjects (Pratt and Atkinson, 2020). Within the discipline of geography, knowledge is scientifically categorised into sets and subsets as shown in Figure 10.1, a structure that is evident in the school geography curriculum, which is defined by the department for education and exam boards.

The coloniality of geography: colonial discourses

Within geography, colonial discourses (Figure 10.2) persist in representations of people and places in, for example, some textbooks and exam specifications (Winter, 2011). Countries such as Nigeria, Kenya, Bangladesh, and India are positioned as underdeveloped or developing, as exotic, undemocratic, and poor, while the United States, the UK, Canada, and Australia are positioned as developed, democratic, and wealthy. These stereotypical discourses distinguish between geographical places on the basis of colonial constructions portrayed as 'facts', rendering invisible the huge disparities that exist *within* each of these countries. In effect, they continue to divide the world into those who are privileged by these distinctions and those who are harmed by them – including the students in our classrooms.

Colonial discourses in education

As discussed earlier, the spaces and places of Euro-Western institutions of power are set up as white possessions, including education which is a white possession both structurally and in the form of the teaching profession, which is disproportionately white. We therefore argue that what takes place in classrooms is affected by the coloniality of the discipline and the curriculum, which are also embedded in the coloniality of the education system (Figure 10.3) (Pirbhai-Illich et al., 2017).

Developed-developing-underdeveloped
Civilised-uncivilised-exotic
Democratic-undemocratic
Superior-inferior
Wealthy-poor
Donor-recipient

FIGURE 10.2 An example of colonial discourses evident in geography education

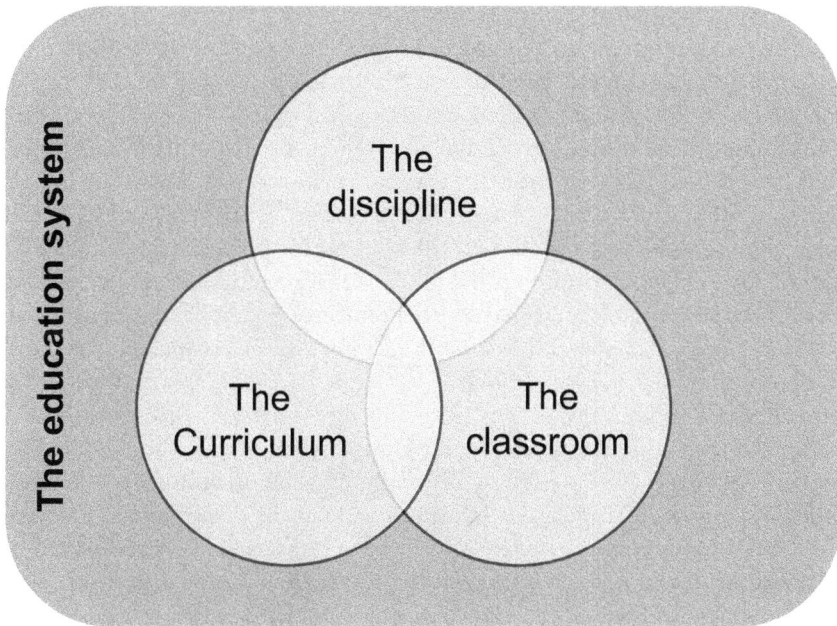

FIGURE 10.3 The coloniality of geography education

The colonial discourses that are evident in mainstream education reflect capitalism's deep entanglement with and in coloniality (Pirbhai-Illich and Martin, 2021).

- *A discourse of commodification* – schools and universities are treated as businesses, placed in competition with each other and measured by 'value for money'.
- *A discourse of universalism* – includes the standardisation of curriculum and assessment using mainstream norms as criteria and the imposition of one-size-fits-all policies.

- *'Othering' discourses* – treating difference as a deviation from the mainstream norm; through the naming and categorisation of differences as Other (SEND, BAME, EAL) which, when combined with *deficit discourses* position racialised students in such categories as a problem to be fixed – failure is individualised (e.g., it's the students' fault, lack of parental support) rather than it being the effect of a system that privileges the mainstream.

The influence of coloniality on practice

In this section, we provide two examples of the influence of coloniality on classroom practice – one from England and one from Canada. In the subsequent section we explore how the same topics might be taught using a de/colonial approach.

Practice example 1: fundamental British values

In 2014, the UK government introduced a legal requirement for schools in England and Wales to promote Fundamental British Values (FBV) (DfE, 2014). The promotion of FBV focuses on four elements: democracy, the rule of law, individual liberty, and mutual respect and acceptance of different faiths and beliefs (DfE, 2014: p. 5). It is expected that curriculum leaders and classroom teachers will meet this requirement through their provision of spiritual, moral, social, and cultural (SMSC) education. In 2020, we (Pirbhai-Illich and Martin, 2020b) argued that the policy constructs the UK as a space and place in which people categorised as 'Black, Asian and Minority Ethnicities' do not belong. During our research we found an image, based on the Union flag, on a commercial website that specialises in posters for schools. The flag is a word graphic and was being promoted as a resource that covers 'everything quintessentially British that would make a fantastic wall display to reinforce British Values in a school' (Charlie Fox Signs, 2021: n.p.). We have found this image on some English school websites where it has been used alongside information about their legal duty to promote FBV. It is a contemporary example of the role of print media in the construction of a national identity, and a critical analysis of the text on the flag reveals who is made visible and who is excluded from this imaginary of British identity. For example, on the flag

- 13 individuals are named, of whom all are white and ten are men.
- There are seven references to the Second World War and the armed forces.
- Cultural references include the Boat Race, Fish and Chips, Steam Trains, Cream Teas, Yorkshire Pudding, football, Eastenders, and Punk.

Amongst other potential contributions, there is no reference to, for example,

- authors such as Andrea Levy and Salmon Rushdie;
- the contribution to the nation of the Gurkhas and the Windrush generation; and

- Cheddar Man (the archaeological find in Cheddar Gorge of the oldest known skeleton who was an African man) or Emma Clark (the first known black woman footballer in Britain).

As an imaginary, it is one that is centred on white, privileged men, and an English (not Irish, Scottish, or Welsh) identity: 'English nationalism, past and present, is the nationalism of an imperial state – one that carries the stamp of its imperial past even when the empire has gone' (Kumar, 2000, p. 577). This resource makes strong implicit statements about who belongs and who does not and if used unquestioningly in the classroom, the ideology behind it will be perpetuated.

Practice example 2: cultural diversity and geographical literacy

In Canada, each province develops its own curriculum. Saskatchewan is a province in the mid-west of Canada with a population of which 70.9% are of European descent, 12.2% are of First Nations or Métis descent, and 11.7% are classified as immigrants[3] or non-permanent residents (Saskatchewan Bureau of Statistics, 2016). Teachers therefore work with diverse groups of students and a key challenge is how to provide each student with relevant, meaningful learning. In the Saskatchewan curriculum, geography is part of social studies (as it is in the broad general phase of education in Scotland – see McKendrick, Chapter 4), along with history and citizenship education. As part of the practice element of a final-year course in a Bachelor of Education programme, Schultz (2015) worked with a Year 9 student to develop his literacy skills through an aspect of the social studies curriculum that requires students to 'interrogate the meaning of culture and the origins of Canadian cultural diversity' (Saskatchewan Ministry of Education, 2009: p. 19). There are seven indicators that students are expected to demonstrate as achievement of this objective, none of which mentions white-settler-Canadians or English as aspects of cultural diversity; the only two groups that are specified are the First Nations, Inuit, and Métis (FNIM) communities and the Francophone community. This portrays a core Canadian national imaginary that is white, English-speaking settler – however, content relevant to the FNIM and Francophone communities are integrated into the social studies curriculum and not provided as an 'add-on'.

More recent immigrants of colour whose cultures and languages are Other to this core are not mentioned at all and thus effectively erased from this representation of Canadian identity. Links to online resources are provided to support the teaching about FNIM cultures with an emphasis on learning about the history of residential schools and treaty education. This focuses on the acquisition of knowledge as an object rather than the teacher starting from an understanding of how each student relates to these histories. In focusing on the harms there is a danger that patronising and saviour discourses will be used which, while important, does not address the contributions that FNIM and non-European immigrant communities have made to Canada.

De/colonising educational relations in geography classrooms

When the identities and histories of students are either not included in the curriculum, or taught in a way that is othering, this creates spaces for educational relationships in which some students belong more than others. Within the field of geography, we have outlined some of the ways in which coloniality and its mechanisms are evident. Specifically, the modern discipline of geography which grew out of the colonial project (de Leeuw and Hunt, 2018; Jazeel, 2019) was instrumental and thus complicit in enabling and reinforcing the controlling mechanisms of colonisation; geographical knowledge categorises, separates, creates binaries, and hierarchies, and these ways of knowing and understanding the world are presented as 'neutral'.

To counter these harmful practices, we propose de/colonising[4] educational relations. The use of the slash indicates that coloniality and decoloniality are in relation – they are deeply entangled, and the slash therefore also signals that our understanding of de/colonising education is not about replacing one tradition with another. The problem is not with European thought itself but with the lack of self-consciousness of its intimate relation to the coloniality of power, which dominates by presenting itself as universal. Santos (2015: p. 192) argues that 'throughout the world, not only are there very diverse forms of knowledge of matter, society, life and spirit, but also many and very diverse concepts of what counts as knowledge and criteria that may be used to validate it'. This diversity of knowledge systems, however, has not gained much recognition in the West, and we see parallels with children's knowledges and ways of researching and expressing themselves geographically (Catling and Martin, 2011).

In our view, de/colonising pedagogies are based on a critical relational understanding of the world, a world that is interrelational, interconnected, and interdependent. Our approaches therefore start from an understanding that teachers are in relation *with* the students, the curriculum, the materiality of schools and classrooms, the community, and the wider world; relations that are entanglements of social, cultural, environmental, historical, and spiritual identities. Central to geographical education is also the relation to land, which is understood as 'foundational to settler colonialism' (Nxumalo and Cedillo 2017: p. 102). A critical relational approach to geography education therefore needs to attend to the ways in which hierarchies of power find expression according to the colonial context. De/colonising geography education is therefore partly about expanding and pluralising the traditions we draw on, of inviting different ways of thinking about the spaces, places, and boundaries of geography and asking questions such as:

- Who decides what counts as geography?
- Whose ways of knowing are included as legitimate geographical thinking and whose are excluded?

But to focus on the discipline in this way – in effect to treat the subject as an object that can be decolonised by the addition of knowledge and viewpoints from marginalised communities – is a project in decanonising rather than decolonising the curriculum. Although useful, we would argue that decanonising a curriculum may still be done through a colonial lens. Further questions to ask ourselves are therefore:

- Whose identities and ways of being and doing are invited into geography classrooms?
- Who decides what gets taught and what gets excluded?
- In what ways do our practices reproduce the colonial dualisms of insider/outsider, superior/inferior, and agentic/helpless?
- How are our responses to all of these questions influenced by our identities?

These are critical, relational questions. Relational thinking is evident in much of geography education but not generally a *critical* relational approach that attends to the hierarchies of power. A critical approach requires teachers to ask: how often do we ask ourselves not only what or how will I teach, but also, who is the self that teaches? How does that quality of my selfhood influence – the way I relate to students, the subjects I teach, the colleagues I work with, and to the world that I exist in?

Practice example 1 revisited: creating critical spaces of belonging

In our work, in conjunction with other concepts, we have successfully used the concepts of invitation and hospitality (Pirbhai-Illich and Martin, 2020a) in conjunction with the geographical concepts of space, place, and boundaries to support teachers in their efforts to de/colonise educational relationships (Pirbhai-Illich and Martin, 2020b). As stated previously, schools and classrooms are not culturally neutral spaces (Kraftl, 2022). They are dominated by mainstream, colonial ways of being and knowing, and it is the teacher's task to make a conscious effort to create these as spaces of democracy and belonging. We recommend a plural and critical intercultural approach that invites the ways of viewing, being, doing, knowing, and valuing of all communities that make up pluralistic Britain. This does not mean rejecting the way of life of the dominant white, mainstream culture that governs how classrooms are organised, the knowledges that inform the curriculum, or the teaching approaches used. It means opening up these spaces for learning to include the knowledges and ways of being of cultures whose identities are seen as different from the mainstream. To do this, we argue that teachers need to first look at their own identities in relation to the socio-historical and geopolitical influences on those identities and how these affect ways in which they might be 'Othering' in their practices. The task of de/colonising therefore starts with oneself, and the questions in Figure 10.4 are designed to support this process.

In our own practice we think of the teacher–student relation as that of host–guest. Teachers have some ownership over their classroom space in that they have

1. Who am I? How do the intersecting dimensions of my identity affect who I am and how I teach? ('race', ethnicity, socio-economic status, religion, gender, ability, sexual orientation)
2. To what extent do my choices over what and how I teach unconsciously reflect my identity? (unconscious bias)
3. To what extent do I consciously invite the ways of being, doing and knowing that are different to mine into the classroom?
4. How can I create spaces of belonging rather than those that are alienating?
5. How can I move away from the divisive (colonial) ways of thinking and being that marginalise and disenfranchise different groups?
6. How can I achieve this without losing the value that can accrue from social classification?

FIGURE 10.4   Questions of identity

the freedom to establish how they and the students relate to each other (Kraftl, 2022; Catling and Pike, 2022). From the students' perspective, they are entering into the teacher's space. If the teacher invites students in such a way that, over time, they develop a sense of shared ownership – this can help foster a sense of belonging for all students. As host, the teacher would not only invite the student but also welcome who the student *is, their* identities, and, by extension, those of their families and the communities and places they are connected to and incorporate these into the curriculum. Therefore, the teacher needs to get to know the students' families, community, and place-based knowledges and incorporate these into the curriculum, which is very different to making curriculum and pedagogical choices based on categorical assumptions about who they are. This may have the power to create potential feelings of belonging, which in turn could counter radicalisation – a key goal behind the duty to promote FBV – by addressing the objectives of democracy and mutual respect and acceptance of different faiths and beliefs.

So, in the teaching of geography, it is also necessary to understand the histories of the students' communities and to ask: how might I teach someone whose community has had negative relationships with my community in the past? How might I teach someone whose community has been forced intentionally and/or unintentionally coerced to assimilate according to my ways rather than theirs? The ways of being/orientations towards working with difference would include inviting plural ways of being, knowing, and doing into the classroom and curriculum spaces; being and learning with and alongside students' different ways of being, knowing, and doing; being attentive to one's own 'translations' of difference and learning about the other with humility, respect, and reciprocity.

We argue that this is a form of hospitality in which the teacher/host identifies *with* the student/guest, not to benefit from that knowledge but to understand her/himself as a learner too. Learning becomes a joint process of exploration, which is responsive to the questions the students have as well as those that are derived from the statutory curriculum. The classroom becomes a space in which all students feel that they can actively participate and thrive as an equal contributing member and where they see their differences reflected and integrated into the curriculum rather than being categorised as 'exotic'.

Practice example 2 revisited: drawing on funds of knowledge

In this example, we demonstrate how Schultz (2015) worked with a Year 9 student of First Nations descent who had been failed by the formal education system and was being educated in an alternative school. The teacher worked with Thomas (pseudonym) once a week for seven weeks. Judged by standardised assessments, Thomas's literacy skills were at the grade 3 level, while his age placed him in grade 8. During the seven weeks the teacher drew on Thomas's funds of knowledge (FoK) to create a culturally responsive programme that would develop his literacy skills through a social studies topic. Funds of knowledge are described as the 'historically accumulated and culturally developed bodies of knowledge and skills essential for household or individual functioning and well-being' (González et al., 2005, p. 72). FoK are not to be confused with prior knowledge or with interests (both of which may not relate to the enduring, everyday cultural and family contexts of the student). By their very nature, FoK are specific to each family and community and put the student in the knowledgeable position rather than the teacher. In this regard, the approach views students' knowledges as an asset rather than deficit in relation to mainstream standards. In their first meeting, the teacher aimed to begin the process of establishing a reciprocal educational relationship with Thomas.

> In our first meeting, Thomas and I shared information about ourselves . . . During our conversation Thomas took out a piece of paper and began drawing detailed outlines of maps from memory, which included the layout of his home town and a map of Saskatchewan. Thomas added routes he travelled with his family. Thomas' funds of knowledge about routes, places and maps came from his home and community. [He had] a vast knowledge of where places were in relation to each other [and] how to get from one to another.
>
> *(Schultz, 2015: p. 24)*

The teacher was then able to build a bridge between Thomas's FoK of mapping and the social studies outcomes (Saskatchewan Ministry of Education, 2009) expected for his grade level:

- IN8.1 Investigate the origins of Canadian cultural diversity; for example, examine the extent to which cultural groups are able to retain their cultural identity in Canada.
- DR8.1 Develop an understanding of the significance of land on the evolution of Canadian identity; for example, illustrate on a map various designated lands in Canada.

The teacher used a multimodal literacies (Kress and Jewitt, 2003) approach to engage Thomas in using print-based, graphic, and geographic literacies to create a map and report that showed the locations of reserves and parks, different types of grain fields, various mines, and wildlife refuse throughout the province. By the end of the programme, he had created two detailed maps and two written texts, each of which acted as authentic assessments of his achievements. Schultz concludes that Thomas's FoK:

> Helped me engage him in reading and writing – two areas Thomas struggled with. Each time we met, he was eager to begin working, and I think he began to regard himself as a capable learner, quickly realising that school can be fun. I hope it created as memorable a learning experience for Thomas as it was one for me.
>
> *(Schultz, 2015: p. 25)*

Conclusion

As we demonstrated at the beginning of the chapter, a critical awareness of one's own identities and the ways in which each of us, in different ways, embody coloniality, is a crucial first step to understanding how this influences our educational relationships. The relationships we develop with students are deeply connected to our identities and their geopolitical and historical locations, and unless we understand how these influence the material, social, and spiritual dimensions of the spaces and places within which geography education is enacted, other forms of decolonisation (e.g., decolonising the curriculum) may be less transformative than intended, may be tokenistic, or may even continue to do harm to the communities that have already been harmed for centuries by the colonial system.

We conclude by arguing that de/colonising educational *relations*, based on a non-hierarchical, non-oppositional, interconnected, interrelational, and interdependent understanding of the world, is a necessary foundation from which it might be possible to de/colonise the curriculum. No space for learning is neutral, whether it is formal or informal. De/colonising educational relations therefore requires teachers to not only ask questions such as what or how will I teach, but they also need to critically engage with the questions we posed in Figure 10.4 that are designed to support the development of a critical awareness of the influence of one's identities on educational relationships.

These questions apply to all teachers/educators, no matter what the discipline or phase of education, but because geography was created through its role as a colonial mechanism there are some specific questions about its structure and how geographical knowledges are constructed that also need consideration. Jazeel (2017) recognises that, for geographers in higher education, there is now 'a collective awareness that geography's decolonial imperative should concern us all, and an equally collective responsibility to think together to do something about the inconvenient truth of coloniality at large in our discipline' (p. 334). In the same issue, Radcliffe (2017) proposes that 'the decolonial turn encourages re-thinking the world *from* Latin America, *from* Africa, *from* Indigenous places and *from* the marginalised academia in the global South, and so on' (p. 329). However, in their focus on knowledge and knowledge production in geography, Jazeel and Radcliffe present decolonisation as an epistemological challenge whereas we would argue that it must also be an ontological one. The demographic of the discipline requires 'geographers to confront white supremacy, white privilege, and racism in the past and present' (Esson et al., 2017) before they can enter into debates about decolonisation and decoloniality, debates which, Esson et al. argue, should be 'determined by those on the margins who have been racialised as Indigenous and non-white by coloniality' (p. 385). Our question is therefore not how can the geography curriculum be de/colonised. Rather we ask how can the relationships that contribute to the structures, institutions, and praxis within which geography's disciplinary knowledge is taught and researched be de/colonised? How can we approach the relationships that support our praxis such that we avoid extractive, colonial methodologies and think beyond the classification and categorisation of the world into hierarchies that continue to divide along lines of privilege and subjugation? We firmly believe that attending to the ways we embody coloniality and working with and alongside the Other with humility, based on the principles of invitation and hospitality discussed earlier, and recognising that this is a lifelong journey may provide a productive starting point.

Notes

1   For a detailed account of the difference between object-focused and relational thinking, see Martin (2013) and Martin and Pirbhai-Illich (2017).
2   We show 'race' in scare quotes to indicate that it is a socially and politically constructed concept. However, we continue to use the concept because of its profound psychological and material impacts on people's lives.
3   Immigrants are classified as those not born in Canada, thus people of European descent are more likely to be considered Canadian than non-European immigrants.
4   We use the slash between 'de' and 'colonial' to indicate that there is 'no utopian decolonising space that is separate from colonising spaces because we are all, always already in 'relationship with colonising discourses and materiality' (Bhattacharya, 2018, p. 15).

References

Bhattacharya, K. (2018) Coloring memories and imaginations of "home": Crafting a de/colonizing autoethnography, *Cultural Studies ↔ Critical Methodologies, 18*(1) pp9–15.

Catling, S. & Martin, F. (2011) Contesting powerful knowledge: The primary geography curriculum as an articulation between academic and children's (ethno-) geographies, *Curriculum Journal, 22*(3) pp317–335

Catling, S. & Pike, S. (2022) 'Becoming acquainted: Aspects of diversity in younger children's geographies'. in Hammond, L., Biddulph, M., Catling, S. & McKendrick, J. H. (eds) *Children, Education and Geography: Rethinking Intersections* Abingdon: Routledge.

Charlie Fox Signs (2021) *British Values Union Jack.* www.charliefoxsigns.co.uk/british-values-union-jack accessed July 29th 2021.

de Leeuw, S. & Hunt, S. (2018) Unsettling decolonizing geographies, *Geography Compass*, *12*(7) pp1–14.

de Rugy, M. (2020) *Geography in the Colonial Context. Digital Encyclopedia of European History*, https://ehne.fr/en/encyclopedia/themes/europe-europeans-and-world/colonial-expansion-and-imperialisms/geography-in-colonial-contex accessed August 17th 2021.

DfE (2014) *Promoting Fundamental British Values as Part of SMSC in Schools.* www.gov.uk/government/publications/promoting-fundamental-british-values-through-smsc accessed July 28th 2020.

Esson, J., Noxolo, P., Baxter, R., Daley, P. & Byron, M. (2017) The 2017 RGS-IBG chair's theme: decolonising geographical knowledges, or reproducing coloniality? *Area, 49*(3) pp384–388.

González, N., Moll, L. & Amanti, C. (Eds). (2005) *Funds of Knowledge: Theorizing Practices in Households, Communities and Classrooms.* Mahwah, NJ: Erlbaum.

Grosfoguel, R. (2011) Decolonising post-colonial studies and paradigms of political-economy: Transmodernity, decolonial thinking, and global coloniality, *Transmodernity: Journal of Peripheral Cultural Production of the Luso-Hispanic World, 1*(1) pp1–37.

Hanson, E. (2020) *What are Indian Reserves?* https://indigenousfoundations.arts.ubc.ca/reserves/ accessed August 1st, 2021.

Harris, C. (2004) How did colonialism dispossess? Comments from an edge of empire, *Annals of the Association of American Geographers, 94*(1) pp165–182.

Jazeel, T. (2017) Mainstreaming geography's decolonial imperative, *Transactions of the Institute of British Geographers, 42*(3) pp334–337.

Jazeel, T. (2019) *Postcolonialism.* London: Routledge.

Kraftl, P. (2022) 'Geographies of educational spaces' in Hammond, L., Biddulph, M., Catling, S. & McKendrick, J. H. (eds) *Children, Education and Geography: Rethinking Intersections.* Abingdon: Routledge.

Kress, G. & Jewitt, C. (2003) (Eds.) *Multimodal Literacy.* New York: Peter Lang Publishing.

Kumalo, S. (2021) (Ed) *Decolonisation as Democratisation.* Cape Town: HSRC Press.

Kumar, K. (2000) Nation and empire: English and British national identity in comparative perspective, *Theory and Society, 29*(5) pp575–608.

Ladson-Billings, G. (2009) Just what is critical race theory and what's it doing in a nice field like education? in Taylor, E., Gillborn, D. & Ladson-Billings, G. (eds) *Foundations of Critical Race Theory in Education,* New York: Routledge, pp. 17–36.

Martin, F. & Pirbhai-Illich, F. (2017) Places, spaces and boundaries: A critical look at the relational in geography classrooms. in Catling, S. (ed.) *Reflections on Primary Geography. The 20th Charney Primary Geography Conference.* Charney: Oxfordshire, pp. 140–146.

McEwan, C. (1998) 'Dismantling the Master's House?': Towards a postcolonial geography. *www.jiscmail.ac.uk/cgi-bin/filearea.cgi?LMGT1=CRIT-GEOG-FORUM&a=get&f=/mce-wan.htm* accessed November 27th 2016.

Moreton-Robinson, A. (2015) *The White Possessive: Property, Power and Indigenous Sovereignty.* Minneapolis: University of Minnesota Press.

Nxumalo, F. & Cedillo, S. (2017) Decolonizing place in early childhood studies: Thinking with Indigenous onto-epistemologies and Black feminist geographies, *Global Studies of Childhood, 7*(2), pp99–112.

Qazimi, S. (2014) Sense of place and place identity. *European Journal of Social Sciences, Education and Research, 1*(1), pp306–311.

Okolosie, L. (2020) 'White guilt on its own won't fix racism': decolonising Britain's schools. *The Guardian.* www.theguardian.com/education/2020/jun/10/white-guilt-on-its-own-wont-fix-racism-decolonising-britains-schools accessed March 30th, 2021.

Pirbhai-Illich, F. & Martin, F. (2020a) Decolonizing the Education relationship: Working with invitation and hospitality, *Critical Questions in Education, 11*(1), pp73–91.

Pirbhai-Illich, F. & Martin, F. (2020b) Fundamental British values: Geography's contribution to teaching about difference. *Primary Geography, 103*, pp23–25.

Pirbhai-Illich, F. & Martin, F. (2021) Beyond possession: De/colonising the education relationship in higher education. in Siseko, K. (ed.) *Decolonisation as Democratisation*. Cape Town: HSRC Press, pp. 80–107.

Pirbhai-Illich, F., Pete, S. & Martin, F. (2017) (Eds.) *Culturally Responsive Pedagogy: Working Towards Decolonization, Indigeneity, and Interculturality*. London: Palgrave McMillan

Porter, L. & Yiftachel, O. (2019) Urbanizing settler-colonial studies: introduction to the special issue, *Settler Colonial Studies, 9*(2), pp177–186.

Prasad, R. (2004) Countryside retreat. *The Guardian.* www.theguardian.com/society/2004/jan/28/raceintheuk.raceequality accessed January 28th 2004.

Pratt, K. & Atkinson, R. (2020) How do some primary schools in England organise and implement the broader curriculum?, *Education 3–13, 48*(4) pp357–364.

Quijano, A. & Ennis, M. (2000) Coloniality of power, Eurocentrism, and Latin America. *Nepantla: Views from South, 1*(3) pp533–580. www.muse.jhu.edu/article/23906.

Radcliffe, S. (2017) Decolonising geographical knowledges, *Transactions of the Institute of British Geographers, 42*(3) pp329–333.

Santos, B. S. (2015) *Epistemologies of the South. Justice Against Epistemicide*. London: Routledge.

Saskatchewan Bureau of Statistics (2016) *Demography and census reports and statistics.* www.saskatchewan.ca/government/government-data/bureau-of-statistics/population-and-census accessed August 16th 2021.

Saskatchewan Ministry of Education (2009) *Social Studies 8.* Regina: Ministry of Education www.edonline.sk.ca/webapps/moe-curriculum-BB5f208b6da4613/CurriculumHome?id=171

Schultz, C. (2015) Drawing on funds of knowledge, *Primary Geography, 87*, pp24–25

Stanek, M. (2019) Decolonial education and geography, *Geography Compass, 13*(12) pp1–13.

Truth and Reconciliation Commission (2015) *The Final Report of the Truth and Reconciliation Commission of Canada*. Montreal: McGill-Queen's Press-MQUP.

Winter, C. (2011) Curriculum knowledge and justice: Content, competency and concept, *Curriculum Journal, 22*(3) pp337–364.

11

CLIMATE CHANGE EDUCATION

Following the information

Steve Puttick, Paloma Chandrachud, Rahul Chopra,
James Robson, Sanjana Singh, and Isobel Talks

Prologue

There are many big stories being told about climate change. Stories of origins, natures, beings, processes, and futures. Stories spanning geologic time and drawing lines to connect people, consumption, technologies, injustices, oppressions, hopes, and fears. Origin stories being told about the Anthropocene represent one aspect of the competition to reposition the present moment in relation to what went before. As a concept, the Anthropocene channels and owns some of these discourses, focusing attention on questions about the ways in which human interactions on a planetary scale are reshaping the very planet itself and about how education should respond. As a concept, the Anthropocene also represents the racialised ways through which so many stories continue to be informed by – and innocent to – extractive colonial logics: 'In its brief tenure, the Anthropocene has metamorphosed. It has been taken up in the world, purposed, and put to work as a conceptual grab, materialist history, and cautionary tale of planetary predicament' (Yusoff, 2018: p. 1–2).[1]

All science is told through narrative to some extent, and this is amplified in the case of climate change. Whose stories are told and how they are told are vital questions (Archibald et al., 2019; Facer, 2019; Mathur, 2017). The present moment is in many ways being told as the climax: the tense hinge point from which one of multiple possible futures will be chosen. Maybe this is always how the present functions for those experiencing and constructing it. The deeply storied nature of climate change adds to the rationale for adopting the approach of fictionalised story in this analysis of children's geographies and their relation to climate change, inspired by the way in which McKittrick (2021) describes and plays with stories, highlighting that 'telling, sharing, listening to, and hearing stories are relational and interdisciplinary acts that are animated by all sorts of people, places, narrative

DOI: 10.4324/9781003248538-15

devices, theoretical queries, plots' (p. 6). The short story we have tried to tell in this chapter emerged through collaborations on the projects *Climate Change Education: Information for Teaching in a Digital Age* (Nuffield), *Climate Change Education Futures in India* (GCRF), and TROP ICSU (Trans-disciplinary Research Oriented Pedagogy for Improving Climate Studies and Understanding; Chopra et al. (2019)). The interdisciplinarity of these particular projects spans education, geography, and climate science.

These collaborations focus on climate change because of its importance globally and in the particular context of teaching and teacher education. Knowledge about climate change has and is rapidly developing, which presents challenges for teachers' professional development, education systems, and curricula. Urgent calls for curricular reform that gives sufficient attention to climate change are being made by young people, academics, policymakers, and others. The recency of climate change knowledge, combined with the dominance of digitally mediated information, makes attention to the digital seeking of information a timely and important task for research. Part of the research this chapter presents is based on a large-scale survey and in-depth interviews with teachers in a range of settings and phases of education in India. The people and places animating this story focus on fictionalised six-year-old children in Pune (India) and Oxford (England), their teachers,[2] and futures. To help the flow of the story we have made liberal use of endnotes, which we hope provide a useful source of information and references to follow further.

Two classrooms

18.5204° N, 73.8567° E

It's springtime, the weather is heating up, and the AC in Saanvi's mum's car is always too cold. That's what Saanvi thinks anyway, and so she wraps herself in her mum's pashmina and looks out at the morning traffic moving slowly by her window. Well, some of it is moving quickly but not always in the same direction as the rest of the traffic. Bikes, auto-rickshaws, well-laden trucks, Ubers, and cars weave across the city, some hiding other schoolchildren behind tinted glass in climate-controlled spaces, others stacked with siblings on a single moped. All making their way from somewhere to somewhere else. The road shines in the heat, and a haze of pollution sits in the air. At the school gates Saanvi leaves the pashmina on the back seat, waves goodbye to her mum, and joins her friends heading into their classroom. After a brief walk across the playground in the warm morning air, the school's solar-powered AC is refreshing to feel and breathe in. Sitting at her desk, Saanvi takes out her books and stationery and looks at the board where Mrs Gupta is writing the title and getting ready to show a short video: *CAFOD, One Climate One World*.[3]

At the end of the school day – 3:30 p.m. in Pune – Saanvi's mum is waiting near the school gates, ready with a cold bottle of water and some laddus.

51.7520° N, 1.2577° W

It's 10 a.m. in Oxford, and Oliver's shoes are still annoyingly wet from the walk to school. He can feel his toes squelching. It looked like it had been raining all night, and it was definitely a bad idea to jump in that puddle. On the Smartboard he can see Miss Smith googling *climate change primary school YouTube*, then clicking on the first hit, a video: *CAFOD, One Climate One World*.[4] A child's voice begins *what is the weather like today?* 'I can tell you that' thinks Oliver: 'rainy. It's always rainy!' *Is the sun shining?* the child's voice goes on. 'No'. *Is it raining?* 'Yes!' Underneath his table Oliver wiggles his toes and pulls his feet out of his shoes to try and get them to dry off a bit. The wet patch under his desk will be there for a while, despite the massive iron radiators on the walls at the side of the room always burning hot. Normally, too hot to touch in the mornings. The windows are open above them, so the classroom doesn't overheat, but there is still condensation on the windows. Miss Smith always *tuts* when she feels the radiators on during warm days. Says it's heating the world up and wasting money. But today isn't warm, and Oliver's attention switches back to the child's voice . . . *this is causing the polar ice caps to melt. This is causing our sea levels to rise, which is bad news for people living near the coast.* 'Phew! Glad I don't live near the coast', Oliver thinks, as the video shows an animation of a man jumping from his nearly flooded house into a small boat and then speeding off into the distance.

Two teachers

The cyclones in the Arabian sea, off the west coast of India, seem to be getting more frequent, and Mrs Gupta is concerned her city – Pune – is going to be increasingly affected. Flooding, high winds, trees falling. What kind of future will her students inherit? Climate change wasn't on the curriculum when she was a student, either at school or during her B.Ed. course.[5] While she didn't learn about climate change during her own formal education, she now often hears about the issue on the news: *what will happen with climate change?*[6] is a question asked across India,[7] with impacts and responses varying from those faced in the north – the Himalayas and its foothills, glaciers, peoples, and other animals,[8] to cyclones and flooding threatening cities bordering the Arabian Sea on the west and the Bay of Bengal on the east, and inland areas increasingly facing periods of extreme heat.[9] It is the need of the hour.[10] Mrs Gupta also hears about climate change from some of her students; Saanvi's older sister is part of the youth-led climate strikes in the city,[11] raising awareness of climate issues and calling for greater attention to be given across policy and curriculum and concrete actions supporting decarbonisation to be taken. Net Zero 2050. It feels like it is a long way from the classroom to their strikes. The times work on different scales. There is an urgency and immediacy in the youth strikers' energy. We needed this done yesterday. We must act now! Time-scales that work differently to the progression of learning planned out over lessons, terms, years, and phases of formal education. What will happen with climate

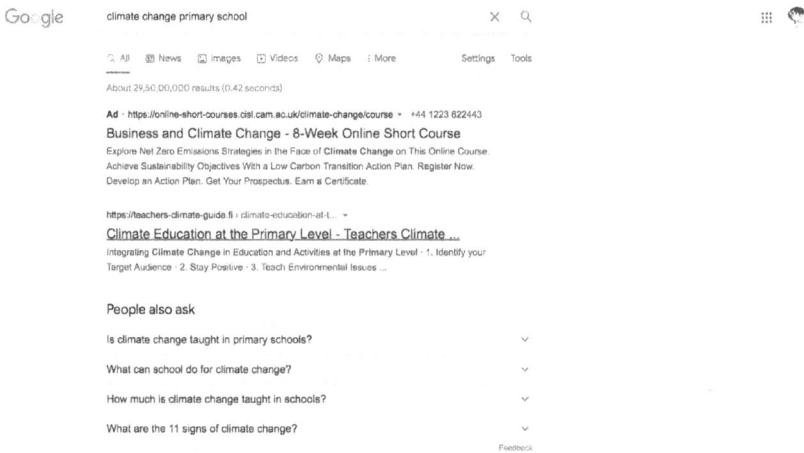

FIGURE 11.1 Screenshot of Google search results for 'climate change primary school', conducted on 19 May 2021 in Pune, India

change over these periods of time? Change. Change makes it all harder. It's all changing, and there is so little time to prepare lessons, mark books, teach, breathe, and maybe eat. Keeping up to date with the latest science about climate change and then deciding what children should be taught now. These are big, complex questions. But so little time.

In just 0.42 seconds Google returns Mrs Gupta with *about 29,50,00,000 results* (Figure 11.1).

The first link is a paid advert: an eight-week online course from the University of Cambridge: *Business and Climate Change*. The second is a Finnish website: *Teacher's Climate Guide*. Underneath the suggestions (*people also ask*) she finds: Campaign Against Climate Change, a UK-based campaigning group, which 'exists to push for the urgent and radical action we need to prevent the catastrophic destabilisation of global climate'.[12] Then, the NPR (National Public Radio) – a non-profit media organisation based in Washington, DC – website and their top tips for teaching climate change: *8 Ways to Teach Climate Change in Almost Any Classroom*.[13] Next down the list – she is starting to scroll to see these results – is the World Wildlife Fund's (WWF– a UK-based charity with headquarters in Woking) climate change resources for schools.[14] The website of a primary school in Doncaster, England, is next,[15] with summary responses to questions including, what is climate change? What is the difference between weather and climate? A NASA (National Aeronautics and Space Administration) Climate Kids video is embedded lower down the page explaining the difference between weather and climate, and beneath this is a CAFOD video (the same one Saanvi and Oliver watched). A PDF resource guide from UN CC:Learn[16] (United National Climate Change Learning Partnership) is the next result; number seven out of *about 295,000,000*. UN CC:Learn is supported by the Swiss Agency for Development and Cooperation. The eighth result is also a PDF – a *Classroom Resource Booklet* – from Climate Change, Science Foundation Ireland.[17] A Routledge book *Teaching Climate Change in Primary Schools*, edited by an Irish academic is next, available for pre-order; Hardback £120, Paperback £24.99.[18] Tenth, a page on the European Union co-funded website globalschools.education[19] which describes and links to a 12-page PDF resource *Teaching About Climate Change in Irish Primary Schools*.[20] The power of the algorithm, and the knowledge and resources to exploit the algorithm: the top ten results from a search conducted in India exclusively deliver websites produced in and representing Euro-American knowledge production[21]; even the website of the single primary school in Doncaster.[22]

Miss Smith, also short on time and searching for inspiration to use in her lessons, puts the same search terms *climate change primary school* into Google. In 0.55 seconds she receives *about 261,000,000 results* (Figure 11.2). The first hit is a link to climate change resources for schools from the WWF (Mrs Gupta's fifth result), which the results page says she has previously visited three times, the last time just over a year ago.[23]

The second result is the primary school in Doncaster, England; Mrs Gupta's sixth result. Third is the same as Mrs Gupta's: Campaign Against Climate Change. Fourth is the BBC (British Broadcasting Corporation): a primary school assembly describing the causes and consequences of climate change.[24] Next is Twinkl, a commercial website. This Climate Change – KS1 Environmental Primary Resources

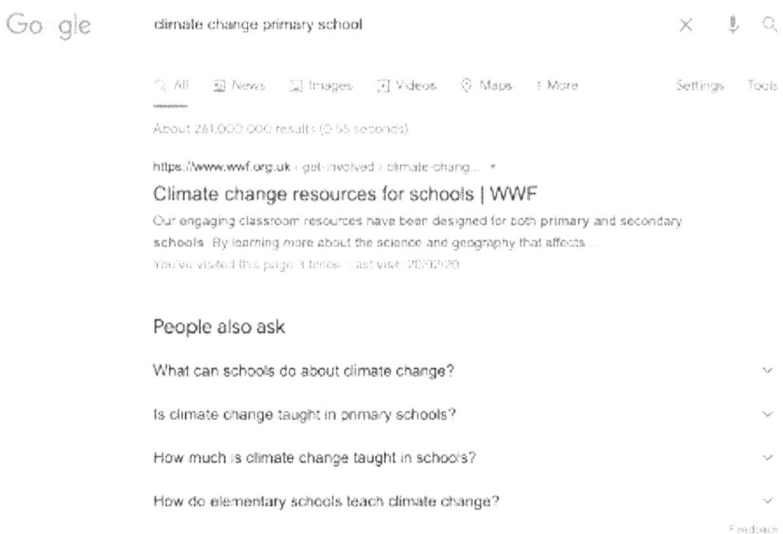

FIGURE 11.2   Screenshot of Google search results for 'climate change primary school', conducted on 19 May 2021 in Oxford, UK

page[25] links to: Climate Change Wordsearch; Renewable and Non-Renewable Energy Sorting Cards; KS1 Greta Thunberg Differentiated Reading Comprehension Activity. The following result is SEEd (Sustainability and Environmental Education): 'It's a scandal how little support and encouragement our education system gives to young people and their learning about sustainability . . . MAKE A DONATION[26]'. The Climate Change page[27] is a list of hyperlinks to resources from UK-based professional associations; Association for Science Education; Geographical Association; Royal Geographical Society with IBG. A Best Practice Article from headteacher-update.com[28] is next, followed by Climate Change for Kids – 16 of the best KS2[29] global warming teaching resources[30]: 'Learn all about Greta Thunberg, Extinction Rebellion, the climate crisis protests, carbon emissions and more with these worksheets, lesson plans and other resources for primary pupils'. The first teaching resource on this page links to another commercial website. The ninth link is to STEM Learning Ltd., whose aim is 'to provide a world-leading STEM education for all young people across the UK'.[31] The first link on their Practical Action: Climate Change page is a free set of PDFs, PowerPoints, and an accompanying video on a *Beat the Flood STEM Challenge*, in which 'students use their STEM skills to help them design and build a model of a flood-proof house'.[32] The tenth result Miss Smith received is the same Routledge book Mrs Gupta received as her ninth result. Below this is a paid advert for a 6-week online course with the Business School at the University of Oxford: *Climate Emergency Programme – Study Online with Oxford.*[33] Where to start? 'The CAFOD video that the other primary school had on their page looked good – I could use that at the start to introduce the topic and get students thinking about' thinks Miss Smith, 'and the WWF resource was great last year; I wonder if there are still copies of those worksheets in the cupboard?'

Journey of a video

The CAFOD video shown in both of these classrooms has a virtual presence embedded on multiple webpages, surrounded by other text, images, and messages. It is stored physically on servers,[34] in quite a different existence to the worksheets in Miss Smith's cupboard: is it a thing? Through interfaces[35] – here, the projectors and interactive whiteboards on classroom walls and speakers, it materialises in the classrooms. Sounds, images, words, and light. Ideas and messages that (might have) once been spoken of in CAFOD's South London head office. Words and pictures drawn on flip charts. Brainstormed. Something to show the world is being heated up – how can we show *the greenhouse effect*? Use a greenhouse? A blanket. Blankets keep the heat in.[36] What impacts do we want to show? What was in the pre-reading?[37] My mate shared something interesting on Facebook this morning – look at this on my phone. What other videos are out there? Search YouTube. Search Climate Change impacts on Wikipedia. Sea level rise. Let's make the house flood. What happens to the guy? He can't just drown – it's for kids! How about a speedboat? Jump into a speedboat and speed away to somewhere else higher, drier, safer. Ideas, coffee, discussions, then decisions fixed onto these images, these words, and these sounds. Blankets. Boats. Who was in the room? Who wrote on the board? Who suggested the ideas that won out? How did those ideas interact with other work and networks of experiences, expertise, and discourses? Moving back from that room, what were their experiences of climate change in school? At university? On the radio during the commute to the office that morning?[38] What did the billboards say? How did they react, and in what ways did this broad milieu inform the discussions and decisions that day? What emails, WhatsApp messages, and phone calls further refined the ideas and shaped the final cut? The networks explode backwards from that meeting room; trajectories, discourses, beliefs, and agendas focused into that one room – friction through debate and decisions – and then crystallised into *the* video. One world. Uploaded in 2014. Then exploded outwards again; downloaded, embedded, shared, and watched hundreds of thousands of times, from Pune to Oxford.

18.5204° N, 73.8567° E

In Pune, Mrs Gupta has found a worksheet on Google images: 'My ideas to tackle climate change'. It's designed to help children think about different things that can be done at an individual, national, and international level by asking 'what can I do', 'what can the government do', and 'what can the world do'. She hits the download button, not really noticing who has produced the sheet.

As she scans other potential resources on her computer screen, Mrs Gupta gradually becomes aware of the discussion on the radio that she has on in the background. Somebody is talking about the ways in which the nature of work is changing, how automation, digitalisation, and artificial intelligence are all changing the jobs that people are doing and will do in the future. The discussion moves on to focus on the skills young people will need in this rapidly changing labour market,[39] with one of the radio guests emphasising the importance of green skills.[40] Mrs Gupta stops what she's doing and, as an impulse, types 'Climate Change Education Future of Work' into Google.

One of the first results she sees is an article on the International Labour Organisation (ILO) website: 'Invest in climate education to build a better workforce for a greener future'.[41] The authors argue that climate change will alter the structure of employment and could lead to the loss of 80 million jobs by 2030, with countries in the Global South hit hardest. The article goes on to highlight climate change education as being a critical tool in helping young people to develop the skills needed for a low-carbon labour market and to help them reimagine their employment futures in ways that will help a just transition to a greener economy.

Mrs Gupta pauses and reflects on the potential her lessons have to empower young people as economic agents of change. She moves the worksheet she had previously downloaded, opens up a blank word document, then clicks on the next link in her search – 'Envisioning the Future of Education and Jobs'.[42]

51.7520° N, 1.2577° W

In Oxford, Oliver's mum is stuck in traffic on the ring road . . . again. She and Oliver are on their way home from the supermarket. Oliver stares blankly out of the rain-spattered back window at the motionless cars all around, barely noticing the cyclists flashing by on the cycle track next to the road. His mum reaches over and turns on the radio. The show host mentions a phrase that Oliver had heard earlier on today at school: 'we must build back better'. What did it mean? On the radio they were talking about 'green jobs' and 'green skills', but he wasn't really sure what those were.

Another cyclist flashed passed his window, throwing up spray from her back wheel. Oliver started thinking about what he'd like to do when he grew up. People often asked him but he never really knew what to say. He still got the same lecture whatever he said: 'work hard at school so you can go to university and then get a good job'. But he wasn't sure what a good job was. He was pretty sure he didn't want a job that meant he had to sit in a car stuck in traffic every day; apart from being such a waste of time, hadn't they just learned how the emissions from cars contribute to climate change.

In fact, the more he thought about it, the more he realised that a lot of the jobs people talked about and that his friends' parents did weren't very good for the environment. He wondered what kind of work would be good for the environment; what kind of work would make the world a better place. He hoped they'd talk about it in school sometime.

Two futures?

So what happens next? What kinds of futures do Saanvi and Oliver choose? Is it like the *Choose Your Own Adventure* book: what future would you like? For Global Net Zero by 2050, turn to page 23. For persistently higher carbon emissions, turn to page 41. In our research with teachers in India, future scenarios were often described with impending senses of urgency and severity:

> Because if I say that there is a burning of these fossils and permafrost is thaw-ing, we will talk about it like it is a bomb that is hidden at the moment and can explode any time. But how that incremental change in temperature every year is going to trigger that and then it will release carbon dioxide, methane into the atmosphere whose effects would be catastrophic. Can't even imagine how much impact it will cause and what will happen to the life on earth.
>
> *(Vamil)*

We can't even imagine how much impact it will cause. Not being able to imagine is connected a wide range of issues about knowledge production, teacher expertise, the use of metaphors, and public discourse about climate change. In this chapter we have tried to imagine aspects related to information. The centrality of online, digitally mediated information means that further work on teachers' searches for information might also draw on critical interventions made in relation to the use of mobile and computer applications more broadly (Graham, 2020; Rose et al., 2020) and particularly their calls to regulate, replicate, and resist hegemonic information capture and circulation. The technologies making information available to teachers are neither 'neutral' nor designed *for* teachers. They are designed to drive clicks, gain attention, and generate money through advertising. In what ways might geography education research empower teachers and contribute to resisting the more malign impacts of these technologies? How can work on anti-racism (Puttick and Murrey, 2020) and worlding (Müller, 2021) of geography expand, decolonise, and diversify the ways in which knowledge about climate change is produced and through which possible futures are imagined and brought into being? We are arguing for climate change education that is connected – including at the level of sources of information – to concerns about epistemic justice and wider, more inclusive and more critical debates[43] because what and whose information is privileged in formal education is central to these arguments. The globally limited nature of current circulations and representations of knowledge about climate change, illustrated through the example of Google searches carried out in Oxford and Pune, is striking: the dominance of the Global North is highlighted by the almost surreal ways in which Irish primary schools feature at multiple points on the Indian search results – and even a single primary school in Doncaster – and yet not a single Indian resource or website makes the first page of results when the search is conducted in India, echoing findings from our recent interviews with teachers in India: videos on YouTube may be 'interesting to watch but only the thing that they are referring to developed countries like America. They need to give more knowledge about the Indian scenario' (Tanirika).

The choices facing young people are serious and exciting. There is great potential for imagining and bringing into being more sustainable and equitable futures: futures in which privileged homeowners do not choose to jump into their speedboats and escape to less vulnerable locations but are instead motivated to act in the interests of the poorest and most vulnerable. Education's role in creating alternative futures has to date been an under-realised resource. Teachers' engagement in the potential for their subjects to play a role in reimagining the future of work and employment has scope to expand and revolutionise the existing relationship of supplying the needs of industry. Parallel to, and complementing Castree et al.'s (2021) case for new models of global environmental assessments, the ambition for education about climate change we are arguing for will:

> Broach normative questions explicitly . . . recognise that 'environmental' issues are entangled with non-environmental ones . . . introduce arguments and justifications (giving them parity of esteem with factual knowledge) . . . seek to build understanding of solutions that might be deemed necessary, possible or desirable.

(p. 65)

However, we are not underestimating the scale of this ambition or the challenges that it raises for teachers' professional development. The limited attention given to subject-specific knowledge in primary teacher education is problematic for all areas, and these issues are exacerbated for the topic of climate change because of the rapidly evolving nature of the subject, its complexity, interdisciplinarity, and the contested nature of online spaces through which information about climate change is mediated. There is an urgent need for teachers and children (from an early age) to be questioning about the superabundance of information available to them and for education systems to support critical engagement with information for teaching. Understanding and responding to the climate emergency is vital for children from primary school age, and Tanirika's point about the need for place-specific information raises timely questions about where and who produces knowledge and for whom. The climate emergency brings these questions into sharp relief because of the ongoing impacts of past injustices and the unequal distribution of risk- and climate-related vulnerabilities.

The complexity of climate change knowledge means that it would be unreasonable to place the responsibility on teachers alone. For different reasons, national- and international-level structures and institutions also have inherent weaknesses for the task of developing local, place-specific, and place-relevant information about climate change. Addressing this – in part, spatial – epistemological challenge should be met more often by relationships between local universities and schoolteachers. This relationship is not just one-way: children's perceptions, experiences, and knowledges about climate change also ought to play a greater role in deepening our understandings of climate change. Children's geographies ought to 'speak back' to and help shape academic attention to climate change. Similarly, where teachers' engagement with climate change knowledge currently positions them as receivers and consumers of this knowledge, their (even 'non-specialist') understandings of curricular decisions, challenges, and solutions have much to offer climate change research because of the ways in which the latter is seeking to speak to and with wider publics. Local collaborations across school/university interfaces may engage teachers with knowledge production and in the process enrich the production, such as through reshaping research questions and enquiries with greater relevance to teachers and children. These relationships offer exciting potential for richer and more meaningful, place-specific collaborations that produce robust, critical knowledge that makes genuine connections with children's lived experiences and helps to build authentic, transformational climate change education that is fit for the challenges facing us and plays a substantial role in not only *understanding* possible futures but *bringing into being* more sustainable, equitable, and just futures.

Notes

1 See also Gilroy's (2018) critique, for example: 'the concept of the Anthropocene seems to be most potent and seductive where history is rendered in its thinnest forms and where the shift into a geological temporality seems unexceptional and obvious' (p. 4).

2 Their teachers do not count themselves as geography specialists but have studied some geography in their own formal education, including at secondary school and, briefly, through their teacher education although the attention given to geography subject-specific issues is limited. In Catling's (2017) terms, there is 'not nearly enough geography!'

3 www.youtube.com/watch?v=v8unGCTWUWI The first hit from a Google search ('climate change primary school youtube') conducted in Pune at the time of writing.

4 www.youtube.com/watch?v=v8unGCTWUWI also the first hit from a Google search conducted in Oxford at the time of writing.

5 For example, see the discussion of environmental education in teacher education in India during the 2000s which, similar to many curriculum policy positions internationally both at the time and in many cases continuing to the time of writing, is now notable for the ways in which climate or climate change are not mentioned explicitly but only implied under the broader label 'environmental education' (Ravindranath, 2007).

6 This question is foregrounded in Wester et al.'s (2019) edited volume *The Hindu Kush Himalaya Assessment: Mountains, Climate Change, Sustainability and People*. See also analysis of the policy dimensions of this and similar questions analysed by Dubash et al. (2018).

7 And increasingly so, as challenge from the Anthropocene 'not only to the arts and humanities, but also to our commonsense understandings and beyond that to contemporary culture in general' (Ghosh, 2016: p. 9) is met in the now widespread usage of climate change, with noticeable shifts even from 2017 to the current time of writing four years later.

8 For example, see Mathur (2015, 2017).

9 For example, see Patel (2018).

10 The phrase 'it is the need of the hour' was used frequently by teachers across India in our interviews to describe the urgency and importance of climate change and responses through education.

11 The international network of Fridays For Future often promotes activity and messages through social media, the Pune group's Instagram is at: www.instagram.com/punefff/

12 http://campaigncc.org/aboutus/missionstatement

13 www.npr.org/2019/04/25/716359470/eight-ways-to-teach-climate-change-in-almost-any-classroom

14 www.wwf.org.uk/get-involved/schools/resources/climate-change-resources

15 www.willowprimaryschool.co.uk/climate-change/

16 www.uncclearn.org/wp-content/uploads/library/resource_guide_on_integrating_cc_in_education_primary_and_secondary_level.pdf

17 www.sfi.ie/__uuid/c650af41-58c9-4c9c-a2f4-969298b860b6/SW-2019-Primary-School-Booklet-(2).pdf

18 www.routledge.com/Teaching-Climate-Change-in-Primary-Schools-An-Interdisciplinary-Approach/Dolan/p/book/9780367631680

19 www.globalschools.education/Activities/Educational-tools/Teaching-about-Climate-Change-in-primary-schools2

20 www.trocaire.org/sites/default/files/resources/edu/teaching_about_climate_change.pdf

21 For further discussion of this point in relation to algorithms of search and information, see Noble (2018), Graham and Zook (2013), and in terms of broader critique of the Anglo-American dominance of disciplinary knowledge production, for example, see Müller (2021), and Marginson (2021).

22 The dominance of Anglo-American knowledge production, here illustrated in terms of the sources of information foregrounded by the search engine results, are echoed in terms of academic knowledge production about climate change education (Puttick and Talks, 2022).

23 Returning to the same topic on an annual basis may be common among teachers' searches for information because of the way in which school curricula are structured

and repeated, often within a similar structure of topics and lessons. For example, see analysis of departmental shared areas in Puttick (2017). There are important implications of these timings and patterns for teachers' professional development and particularly the scheduling of subject knowledge enhancement provision.

24 www.bbc.co.uk/teach/school-radio/assemblies-ks1-ks2-climate-change-global-warming/zbgxjsg
25 www.twinkl.co.uk/resources/topics/eco-recycling-environment/climate-change-eco-recycling-and-environment-topics-key-stage-1
26 https://se-ed.co.uk/edu/
27 https://se-ed.co.uk/edu/climate-change-education-resources/
28 www.headteacher-update.com/best-practice-article/teaching-climate-change/221221
29 KS2 (Key Stage 2) is the term used in England to refer to the schooling of 7–11-year-olds.
30 www.teachwire.net/news/climate-change-and-global-warming-ks2-teaching-resources
31 www.stem.org.uk/about
32 www.stem.org.uk/elibrary/resource/34167
33 https://onlineprogrammes.sbs.ox.ac.uk/presentations/lp/oxford-climate-emergency-programme
34 For example, see Amoore's (2020) discussion of the physical locations and materiality of the cloud.
35 In particular, see Rose's (2016) argument for the significance of the concepts interface, friction, and networks in understanding and analysing digital objects.
36 Metaphors and images do imaginative and political work, for example, see Castree's (2020) critical discussion of the use of metaphors in ongoing discourse.
37 For example, research papers published by NGOs like CAFOD: https://learn.tearfund.org/-/media/learn/resources/reports/2020-tearfund-consortium-powering-past-oil-and-gas-en.pdf
38 In particular, see Manzo and Padfield's (2016) analysis of the very particular ways in which media representations construct ideas about climate change (through a Malaysian context), concluding that

> [C]limate change has been framed as both a multi-scalar responsibility and a positive opportunity for two key stakeholders in development, i.e. neoliberal market forces and geopolitical actors keenly interested in restructuring the international political economy along lines reminiscent of the new international economic order (NIEO) demands of the 1970s.
>
> *(p. 60)*

39 For example, see Frey (2019) and Goldin (2021).
40 'Green skills' are defined by the OECD (2014) as 'the knowledge, abilities, values and attitudes needed to live in, develop and support a sustainable and resource-efficient society'.
41 www.ilo.org/global/about-the-ilo/newsroom/news/WCMS_781859/lang–en/index.htm
42 www.oecd.org/education/Envisioning-the-future-of-education-and-jobs.pdf
43 Including arguments for media literacy based on the rights of the child, for example, see Cannon et al. (2020), and opening up

> possibilities to nurture alternative imaginaries and revolutionary potentials . . . Capacious, fluid, creative, and subversive thinking is necessary not only in further critiquing complexities of empire, imperialism, and capitalism but also decentering them and fostering cognitive and epistemic justice.
>
> *(Sultana, 2021: p. 7)*

References

Amoore, L. (2020) *Cloud Ethics: Algorithms and the Attributes of Ourselves and Others*. London: Duke University Press.

Archibald, J., Xiiem, Q.Q., Lee-Morgan, J.B.J. and De Santolo, J. (Eds.). (2019) *Decolonizing Research: Indigenous Storywork as Methodology*, London: ZED Books.

Cannon, M., Connolly, S. and Parry, R. (2020) Media literacy, curriculum and the rights of the child, *Discourse: Studies in the Cultural Politics of Education*, DOI: 10.1080/01596306.2020.1829551.

Castree, N. (2020) The discourse and reality of carbon dioxide removal: Toward the responsible use of metaphors in post-normal times, *Frontiers in Climate*, 2. Doi:10.3389/fclim.2020.614014.

Castree, N., Bellamy, R. and Osaka, S. (2021) The future of global environmental assessments: Making a case for fundamental change, *The Anthropocene Review*, 8(1) pp56–82.

Catling, S. (2017) Not nearly enough geography! University provision for England's pre-service primary teachers, *Journal of Geography in Higher Education*, 41(3) pp434–458.

Chopra, R., Joshi, A., Nagarajan, A., Fomproix, N. & Shashidhara, L.S. (2019) *Climate Change Education Across the Curriculum*, in W. Leal. & S. Hemstock (Eds.) *Climate Change and the Role of Education*, Cham: Springer, Chapter 4.

Cook, I. et al. (2017) From 'Follow the thing: Papaya' to followthethings.com, *Journal of Consumer Ethics*, 1(1) pp22–29.

Dubash, N.Z., Khosla, R., Kelkar, U. and Lele, S. (2018) India and climate change: Evolving ideas and increasing policy engagement, *Annual Review of Environment and Resources*, 43, pp395–424.

Facer, K. (2019) Storytelling in troubled times: What is the role for educators in the deep crises of the 21st century? *Literacy*, 53(1) pp3–13.

Frey, C. B. (2019) *The Technology Trap: Capital, Labor, and Power in the Age of Automation*. Princeton: Princeton University Press.

Ghosh, A. (2016) *The Great Derangement: Climate Change and the Unthinkable*. London: University of Chicago Press.

Gilroy, P. (2018) "Where every breeze speaks of courage and liberty": Offshore Humanism and Marine Xenology, or, Racism and the Problem of Critique at Sea Level, *Antipode*, 50(1) pp3–22.

Goldin, I. (2021) *Rescue: From Global Crisis to a Better World*. London: Hodder and Stoughton.

Graham, M. (2020) Regulate, replicate, and resist – the conjunctural geographies of platform urbanism, *Urban Geography*, 41(3) pp453–457.

Graham, M. and Zook, M. (2013) Augmented realities and uneven geographies: Exploring the geo-linguistic contours of the web. *Environment and Planning A*, 45(1) pp77–99.

Manzo, K. and Padfield, R. (2016) Palm oil not polar bears: Climate change and development in Malaysian media, *Transactions of the Institute of British Geographers*, 41(4) pp460–476.

Marginson, S. (2021) *Globalisation: The Good, the Bad and the Ugly*, Centre for Global Higher Education, Available at: www.researchcghe.org/publications/working-paper/globalisation-the-good-the-bad-and-the-ugly/

Mathur, N. (2015) "It's a conspiracy theory and climate change" Of beastly encounters and cervine disappearances in Himalayan India, *HAU: Journal of Ethnographic Theory*, 5(1) pp87–111.

Mathur, N. (2017) The task of the climate translator, *Economic and Political Weekly*, 52(31), pp77–84.

McKittrick, K. (2021) *Dear Science and Other Stories*. Durham: Duke University Press.

Müller, M. (2021) Worlding geography: From linguistic privilege to decolonial anywheres, *Progress in Human Geography*, pp1–27. DOI: 10.1177/0309132520979356.

Noble, S. (2018) *Algorithms of Oppression: How Search Engines Reinforce Racism*. New York: New York University Press.

OECD. (2014) *OECD Green Growth Studies: Greener Skills and Jobs Highlights*. OECD. www.oecd.org/cfe/leed/Greener%20skills_Highlights%20WEB.pdf

Patel, S.G. (2018) *Community Level Assessment of Heatwaves in Odisha State, India: Effects, Resilience and Implications*. Research Brief. New Delhi: Population Council. https://knowledgecommons.popcouncil.org/cgi/viewcontent.cgi?article=1477&context=departments_sbsr-pgy

Puttick, S. (2017) 'You'll see that everywhere': Institutional isomorphism in secondary school subject departments, *School Leadership and Management*, 37(1–2) pp61–79.

Puttick, S. and Murrey, A. (2020) Confronting the deafening silence on race in geography education in England: Learning from anti-racist, decolonial and black geographies, *Geography*, 105(3), pp126–134.

Puttick, S. and Talks, I. (2022) Teachers' sources of information about climate change: A scoping review, *The Curriculum Journal*, 33(3), pp378–395.

Ravindranath, M. J. (2007) Environmental education in teacher education in India: Experiences and challenges in the United Nation's Decade of Education for Sustainable Development, *Journal of Education for Teaching*, 33(2) pp191–206.

Rose, G. (2016) Rethinking the geographies of cultural 'objects' through digital technologies: Interface, network and friction, *Progress in Human Geography*, 40(3) pp334–351.

Rose, G., Raghuram, P., Watson, S. and Wigley, E. (2020) Platform urbanism, smartphone applications and valuing data in a smart city. *Transactions of the Institute of British Geographers*, doi:10.1111/tran.12400.

Sultana, F. (2021) Progress report in political ecology II: Conjunctures, crises and critical publics, *Progress in Human Geography*, DOI: 10.1177/03091325211028665.

Wester, P., Mishra, A., Mukherji, A. and Shrestha, A. B. (2019) *The Hindu Kush Himalaya Assessment: Mountains, Climate Change, Sustainability and People*. Cham: Springer. https://library.oapen.org/handle/20.500.12657/22932.

Yusoff, K. (2018) *A Billion Black Anthropocenes or None*. Minneapolis: University of Minnesota Press.

12[1]

EXPANDING STUDENTS' CONCEPT OF 'HOME'

Teaching migration with a geographic capabilities approach

David Mitchell and Tine Béneker

We use two case studies, one from a school in England, and the second from a school in the Netherlands. In both schools, the teachers of geography wanted to challenge an 'official' version of school geography which oversimplified and sanitised representations of migration, place, and interconnection. In the English case, the official version of migration was the exam specification; in the Netherlands case, it was the textbook. In both cases, the concept of 'home' (in relation to globalisation, development, and identities) was used to change and develop lessons on migration. 'Home' was not a concept the teachers had focused on before, but they did so after talking to academic geographers, in a way which connected migration to students' lives, to deepen students' geographical understanding of migration.

Introduction: GeoCapabilities phase 3 – teacher scholarship as social justice

The GeoCapabilities approach is about applying a human capabilities lens to exploring how geographical knowledge can contribute to the capabilities students need to live a life that they value. Since Lambert et al. (2015) argued for its potential, GeoCapabilities has been widely recognised and explored as a concept in the international geography education community. Whilst there is some empirically based research into the potential of a GeoCapabilities approach to developing school geography curricula (Bustin, 2019) in the main, GeoCapabilities has been discussed more as an ideal of the educational potential of geographical knowledge. In Western education systems, many young people are taught in schools where often accountability for exam results limits teachers' curriculum making freedoms (their curriculum agency), which in turn can diminish, in some schools, students' access to geographical knowledge and the capabilities-building it supports (see Mitchell, 2020). We would argue that this creates unequal access to knowledge

DOI: 10.4324/9781003248538-16

and therefore is a social justice issue. Ultimately, the GeoCapabilities approach seeks to refocus the attention in schools away from what has become, de-facto, a narrow purpose of attaining exam results to a more meaningful education or, as Bustin states, shifting the focus of school education from 'outputs to outcomes' (Bustin 2019: p. 161).

Early in the GeoCapabilities 3 project, members of the project team (geography teacher educators from education institutions in universities in England, France, the Netherlands, Czechia, and Belgium) initially interviewed 12 teachers of geography (two each from France, the Netherlands, Czechia, and Belgium and four from England) who had volunteered to participate in the project (referred to as associate teachers). These associate teachers all taught in schools identified as being challenging in that they served socioeconomically deprived neighbourhoods where unemployment was high and student attainment low (Biddulph et al., 2020). The associate teachers were interviewed to get a first impression as to the place of migration in their schools' geography curriculum. In relation to subject content, it was found that in the different participating countries migration is often taught in relation to demography; origins, destinations, and the push-pull model; policies in the European Union (EU); the Mexican–US border; forced migration and migrants in society. The focus was often on the impact of migrants on receiving countries, and migrant populations were frequently presented as a poor minority. In the interviews the associate teachers highlighted some specific issues they faced when teaching, including accessing new developments in the academic discipline and a tendency to 'play safe' in teaching about challenging geographical topics like migration:

> [T]eachers would turn to developments in the academic discipline in order to inform their own understanding if they could find realistic ways of doing so . . .
> . . . the interviews have revealed a tendency to resort to 'safe' curriculum content and 'safe' pedagogical approaches, even though this is unlikely to build more positive 'relational capabilities' between young people and the migrants they learn about.
>
> *(Biddulph et al., 2020: p. 271)*

'Relational capabilities' mentioned here is one of the four key concepts associated with broader notions of social justice that the project used to frame its work. These are drawn from Walker (2006) and relate to the principles of capabilities. The others are agency; distributive justice; and mutuality/misrecognition. Utilising these concepts enabled the project team to consider social justice dimensions and challenges in teachers' work, especially in teaching a potentially controversial issue such as migration (Biddulph et al., 2020). Social justice underpinned the work of the project in a range of ways, including the decision to focus on teaching the topic of migration; the decision to work with schools considered 'challenging'; utilising the concept of capabilities and its association with social justice (Walker, 2006); and finally, through developing associate teachers' understanding of powerful disciplinary knowledge (PDK) and how to utilise it in their teaching.

In this chapter we are focusing on teachers' scholarship as a form of teachers' agency in creating the curriculum they believe to be appropriate for the education of their students. Our notion of scholarship is drawn from Healy (2022), who argues that curriculum making requires teachers 'to be alert to who is (re)shaping geographical knowledge both in the academy and for school geography' (p. 195) and that scholarly teaching requires of teachers 'intellectual endeavour and considered decision-making' (Brooks, 2010, p. 69, cited in Healy, 2022). This approach to teachers' work is, we argue, a form of social justice in terms of shaping what school students learn in school geography.

To consider teachers' curriculum agency, we draw on the concept of 'recontextualisation'. We use Bernstein's notion of the pedagogic device and its field of production, recontextualisation and reproduction (Bernstein, 1990). In the field of recontextualisation, Bernstein refers to the 'official' field of reproduction (often state sanctioned, such as examination requirements and sometimes, textbooks) and the pedagogic field (which may be educators, for instance). In the case studies from England and the Netherlands that we present later in this chapter, we look in particular at how a form of teacher scholarship can challenge a restrictive 'official' curriculum by enabling teachers to more directly access and utilise disciplinary knowledge in their teaching.

The associate teachers in the GeoCapabilities project acted as participants and co-investigators in a form of collaborative action research. They were selected by being known to the partners as innovative members of the geography education research community and they expressed an interest in this collaborative project. This can be described as a small, purposive sample of teachers. We stress this is not representative of all teachers, but, importantly for this project with our social justice focus, they were all serving a relatively low socioeconomic catchment in their schools.

Migration and the concept of 'home' in the curriculum

In addition to interviewing associate geography teachers, at the start of the Geo-Capabilities 3 project, 12 academic geographers from across the five countries whose expertise lies in migration were also interviewed, the focus being about research developments in this field in the past ten years.[2] The academic experts highlighted the dynamic approach to the phenomenon of migration and mobilities (temporary stays and as part-and-parcel of life) in academic research, with transnationalism as a main perspective and the migrant population seen as a majority-minority in neighbourhoods and cities who had agency and who contributed to society.

Whilst the concept of 'home' has been a growing part of geographical research and thinking in migration for over 20 years (Blunt and Dowling, 2006; Fathi, 2021; Boccagni, 2017), it was not identified as part of the (official) geography curricula in any school participating in the project, and it was not used by any of the project teachers as an explicitly taught concept in their school geography.

The Dutch and English teachers, in conversation with both the academic geographers whose research expertise was migration and with geography teacher educators, identified 'home' as a geographical concept which could add value to their teaching. The teachers were enthusiastic about what Nieto (2021) describes as the attention shift (in academic research) from a focus on home as the country of origin to a focus on practices of home-making in the places connected to migration which are tied to questions of belonging. Ideas such as home-making open the door to more humanistic approaches in the content of the school geography curriculum with the potential to focus on everyday and personal geographies, including those of the school students, which as Hammond (2021) argues, are often absent in curricula decisions and choices (Catling and Pike, 2022).

Participants and case studies selection

As stated earlier, the GeoCapabilities 3 project was a collaboration between six university partners (two from England and one from each of the other participating countries) and Eurogeo.[3] All university partners had good connections to schools and teachers and the outcomes of the project are published on the project website.[4] However, in this chapter, we report on the work of two specific case study schools, the first is in Halifax, England, and the second in Utrecht, the Netherlands. In both schools, fully informed consent of the teachers and their headteachers was gained, following BERA (2018) guidelines in the English school. Ethical approval was obtained from IOE, UCL's Faculty of Education and Society. There is no equivalent to BERA in the Netherlands, but the same principle of informed consent was followed. Individual students are not identified in this chapter, but the teachers and schools are happy for real names to be used and their school to be identified. Care was taken to ensure there was no disruption on their prime focus, namely on day-to-day teaching.

For this chapter, we have chosen to focus on three teachers in two different schools that we recruited to the project. These are Daniel Whittall in Trinity Sixth Form Academy, Halifax, England, and Klaas Danhof and Anita Mocking in Gregorius College, Utrecht, the Netherlands. The teachers were working in catchment areas affected either directly or indirectly by de-industrialisation and migration. The reason for choosing these two cases in this chapter is that they illustrate a form of teacher scholarship which challenges the 'official' version of school geography in each country. In the English case, the official version of migration was embedded in the examination specification; in the Netherlands case, it was embedded representation in the textbook. In both cases, the concept of 'home' (in relation to globalisation, development, and identities) was used to change and develop lessons on migration.

Daniel Whittall joined the project as an associate teacher at IOE. He is an experienced teacher and currently teaches 16–18-year-olds on the Advanced Level

(A-Level) Geography course at Trinity Sixth Form Academy in Halifax. Halifax is a town in an old industrial area of northern England, whose textile manufacturing industry has declined and which suffers relatively high levels of socioeconomic deprivation. Halifax changed, demographically, with an influx of migrants, mainly from South Asia (UK commonwealth countries) particularly in the 1950s and 1960s, so many second- and third-generation British Asian families live there. Daniel has a doctorate and has published papers in academic and professional journals of geography education and is also active on social media particularly in respect to anti-racist education and decolonising geography. Clearly, Daniel's activity and level of engagement with the geography education community is unusual, and he was selected in part for his interest and existing engagement with geography education. We recognise this limits the scope of our research, and we do not make claims that the GeoCapabilities approach will necessarily support teacher scholarship for others, in the same way as it has for Daniel. Nonetheless, his case illuminates a form of teacher scholarship which we argue is strongly supported by the GeoCapabilities approach in the project.

Klaas Danhof joined the project as head teacher in a school, St. Gregorius College, which participates in a network with Utrecht University. Klaas has a master's degree in cultural geography and a strong interest in experiential geography education that makes his teaching meaningful for the diverse group of students in his classes. He has been involved in international projects and has published in *Geografie*, a journal for practitioners. Klaas involved Anita Mocking in the project, one of the two other geography teachers in the school. She teaches in lower secondary pre-vocational (mavo) classes at Gregorius. She has a broad international experience, in non-educational sectors as well and returned in 2016 to her 'own' school to teach geography.

St. Gregorius College is a 150-year-old, inner city school in Utrecht. Very recently, the school changed the educational approach into a Steiner School (Vrije School). The school has been under pressure for the last ten years because of poor output performances in high-stake exams, lower intake of students, and having an image problem, being seen in the community as a less desirable school to attend. The school recruits students from different parts of the city, but the majority are second- or third-generation migrants. The grandparents of these students arrived in the Netherlands in the 1960s–1970s as guest labourers working in the industrial sector, mainly from Turkey and North-African countries, especially Morocco. Many students live in diverse neighbourhoods with a relatively low socioeconomic status. The teachers were eager to join the project with students in pre-vocational education (12–16-years-old), as these groups are often overlooked in this type of project.

Methodology – a GeoCapabilities 3 approach

In phase 3 of the GeoCapabilities project, the project team supported the associate teachers in four discernible steps, summarised as a diagram (Figure 12.1).

Current geography
(powerful disciplinary knowledge - PDK)

Step 4 - Evaluation & reflection
Assess & evaluate how geographical
concepts, contexts and procedural
knowledge and skills have developed
from the lessons. Return back
to your vignette, to reflect on how
far and why your original ideas
changed.

Step 1- Thinking Geographically
Update and Engage with academic
Geography... get inspired, write and share
a vignette to clarify your thinking.

1. VIGNETTE

**SOCIAL
JUSTICE**

4. Evaluation

2. PLANNING

3. TEACH LESSONS

Step 3 - Teaching
Teach the lessons .. assessing
what the students know before
the teaching.

Step 2 - Lesson planning
Use the PDK planning tool, leading to assessing
students' current knowledge and to consider
which powerful pedagogies to adopt.
Find a rich resource or curriculum artefact on
which to build a lesson/s.

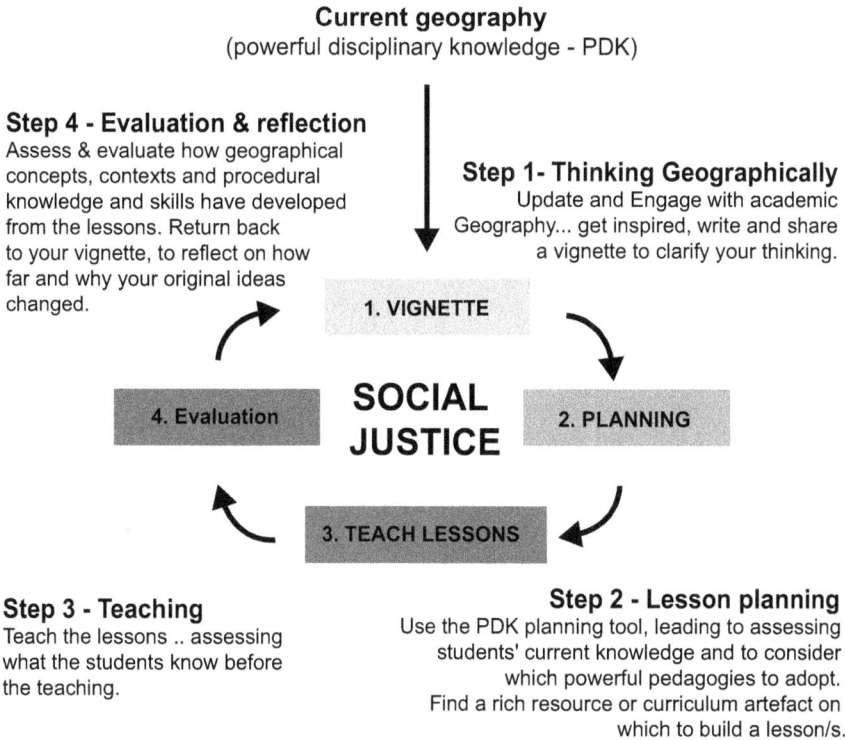

FIGURE 12.1   A GeoCapablities process in practical steps

Step one: engagement with academic geography (seminar/writing a vignette).

Step two: lesson planning/curriculum making (development of a rich 'curriculum artefact' on which to build a lesson sequence). At this point, a preteaching assessment of students' knowledge and understanding of migration was made.

Step 3: teaching the lessons observed by the university partner where possible.

Step 4: evaluation of how knowledge, understanding, and ideas had developed (involving the students and the teacher).

The first step 'thinking geographically' is key to our argument for a form of teacher scholarship that is supported by our project approach. In the English (IOE) case, two academic geographers met with the English partners and associate teachers in a seminar event to share research ideas about migration geographies in university and schools. These were Dr Ben Page, an Associate Professor of Human Geography and African Studies at UCL with research interests including current developments in international migration studies, and Dr Anabelle Wilkins, a postdoctoral researcher at the University of Manchester with research interests including the spatial relationships of Vietnamese diaspora in London. At the seminar, Ben

Page presented and led a discussion about some of the misconceptions as well as trends in international migration (see Castles et al., 2020). Anabelle Wilkins also presented and led a discussion drawing on her doctoral research on the Vietnamese diaspora in London which led to her developing ideas of migrants, work, and home-making in the city (Wilkins, 2019).

In the Dutch (UU) case, a similar seminar was held with two migration experts invited. These were Dr Ilse van Liempt, Associate Professor at Utrecht University, and Dr Joris Schapendonk, Associate Professor at Radboud University. Ilse and Joris presented some of their work on migration trajectories from Africa and through Europe (Schapendonk, 2020) and inclusion and exclusion of young refugees in public space and society (Van Liempt and Staring, 2021). Moreover, they had an open discussion with the teachers and shown a great interest in their practices, concerns, and ideas in relation to teaching migration to their (migrant) students. One important topic was how to address the negative public and political debate around migration and migrants in Dutch society when teaching students with migrant backgrounds and how in geography education sometimes stereotypes are confirmed instead of challenged. The concept of home related to migration was identified as an opportunity.

Teacher educators played a key role in enabling teacher scholarship here. In the English case the academic geographers (Ben and Annabelle) were chosen and invited through personal networks in the university. Whilst they were keen to talk with teachers of geography, it is significant to note that the opportunity for discussion was created by the teacher educator acting as a bridge that connects academic researchers with teachers of geography. In the Dutch case, the meeting with the academic experts was inspiring, but the follow-up was done by the teacher educators – Hans Palings and Tine Béneker, both at Utrecht University. Hans played an important role in translating ideas from academic migration experts into several concrete and specific suggestions and options for the teachers to think about and choose from. Moreover, the Dutch teachers used the conceptual resources (see later) as well. The two teachers designed their lessons together, in close cooperation with the teacher educators. The teacher educators were involved in the evaluation of the teaching, interviewing teachers, and observing the evaluation with students.

Following the seminar event in London, the teachers reflected on the ideas and wrote 'vignettes' to articulate a specific aspect of the PDK of geography. In this chapter, we expand on Daniel's case as an illustration alongside the Netherlands example. But it is worth mentioning that the teachers at the London event developed a range of geographic focuses. These included one teacher who wanted to turn around the notion of migration as always into the UK, to look at emigration out of the UK by using identities and a nuanced conception of 'home' which can cross national borders and be about interconnections, as much as a single location (drawing on Wilkins, 2019). Another of the teachers changed their teaching about migration as a result of the London event. They were influenced by Ben Page's presentation, which drew from Castles et al. (2020) to teach migration as complex, interconnected and as an opportunity as well as (sometimes) a crisis. And Daniel (whose case we expand on later in the chapter) wanted to challenge an 'insider-outsider' mindset to how 'home' is constructed in an area (Halifax) rich in recent migration flows.

Two conceptual resources which the teacher educators and teachers explored together to put ideas into practical planning and evaluation were Maude's typology of powerful geographical knowledge (Maude, 2016) and Klafki's questions about the 'significance' of content taught and learnt (Klafki, 2000). Klafki raises questions on the exemplary, contemporary, and future significance of the chosen content in the case of this project migration. However, the teachers found Klafki's concepts quite abstract and somewhat removed from their day-to-day teaching and lesson planning decisions. They needed more practical ways to help their planning with capabilities and powerful knowledge in mind. Béneker's (2018) fourfold model of powerful knowledge in geography helped the teachers make these practical connections. It was adapted by the teachers into a flexible planning tool which they could then adjust to any particular aspect of migration they were teaching, helping both the teacher (in their curriculum making and evaluation) and the students, who were asked these questions before and after teaching, to make sense of geographical concepts. The tool helps to identify and order key terms and specific knowledge, consider its significance for the students in relation to the world outside, and pays attention to the reliability and trustworthiness of concepts and ideas – asking 'how do we know this?' and 'what sources of information or evidence can we use?' Figure 12.2 shows an example of the tool adapted for this study.

The planning tool was combined with a student focus group and concept mapping exercise with students in a 'post-teaching' evaluation using a common set of questions and key words cards (allowing for some flexibility). The focus groups were deliberately held approximately one month after the completion of the teaching on migration and the planning tool. This was to evaluate deeper learning, after a period of time had elapsed. For the concept mapping exercise, teachers used a common set of concepts students will have learnt about in their lessons on migration (with the flexibility to add additional terms as needed), which students discussed. They were then asked to draw and annotate connecting lines between concepts. To facilitate the discussion with students, the teachers used a set of prearranged questions with flexibility to extend these to prompt and extend the students' talk about migration.

With these tools, teachers were enabled to reflect and think deeply about their teaching of migration. This led to curriculum and pedagogical innovation, and in particular the teachers were able to focus on how their students experienced and thought about ideas and concepts around migration. One such concept is 'home', and we will now explore this through some of the project's case studies of teachers working collaboratively with other teachers and academic geographers.

Findings – two case studies of teacher scholarship – using a concept of 'home'

The following section summarises the case studies[5] which were co-constructed between teacher and geography teacher educators working on the project.

Key terms	Specific geographical knowledge
1. What does the word home mean? 2. What is migration? 3. What is the difference between immigration and emigration?	1. What facts do you know about migration? 2. What countries do people from England migrate to? 3. Why do people make a home in these countries?
What's this got to do with the world outside the classroom? 1. Is emigration an important issue? 2. Should people be able to make their home wherever they choose?	**How do we find out more?** 1. What do you think are reliable or trustworthy sources of information about migration? 2. What do you think are unreliable or untrustworthy sources of information about migration?

FIGURE 12.2 An example of the planning tool developed by teachers on the GeoCapabilities project

Case study 1 – Daniel Whittall, Trinity sixth form academy, Halifax, England

Daniel teaches a geography exam course to 16–18-year-olds including a unit called 'changing places'. The GeoCapabilities project helped Daniel to appreciate that teachers do not have to settle for the particular way that official curricula package geographical knowledge but that teachers have the capacity to enact curricula that are shaped by their own interpretations of powerful knowledge and to contest the received wisdom of the official curriculum.

Daniel was concerned that 'official' (examination course syllabus) encourages teaching 'endogenous' (internal) and 'exogenous' (external) factors that shape places. In the example scheme of work provided by the examination board and in most textbooks and online resources, migrants are categorised as exogenous factors. This forges bonds between a conception of place, exogenous factors, and migration, thus risking differentiating migrant communities from the places they settle in, reinforcing dangerous stereotypes of migrants as externalised 'others'. But a more nuanced teaching of place is possible, as this excerpt from Daniel's vignette shows:

> This sedentarist metaphysics has sometimes led to a bounded conception of place and, in the worst instances, has pathologised human mobility as a 'threat and dysfunction' (Cresswell, 2006: 38) . . . however, it is possible to teach place, and perhaps especially urban places, as assemblages . . . yet nonetheless still shaped by particular structuring processes. One way into this approach is to teach urban places through Massey's (1994) 'global sense of place'. Such an approach highlights how the meanings associated with urban places are derived from the flows that pass through it, from the particular coming-together of specific environmental, social, economic and political processes at a set point in time.
>
> *Excerpt from Daniel's vignette (Whittall, 2021b)*

Daniel drew from academic geography to shape alternative approaches to the concept of place. These included Jazeel (2019), whose relational approach to place takes account of the colonial and racialising legacies of the discipline of geography. Jazeel argues for moving away from bounded conceptions of place that all too easily allow us to classify migrants as externalised 'others'. He deepened his conceptual understanding of place and migration through re-engaging with the discipline and then applied this to teaching case studies of the students' home town and Notting Hill in London. Daniel then carefully selected rich resources that became 'curriculum artefacts' in his teaching. One artefact Daniel felt was particularly powerful for students was a historical photograph taken inside a disused textile factory showing 'no smoking' signs in various languages, including Urdu and Punjabi. This was because the photography helped students to grasp the problem of trying to categorise place-making factors (including people) as either inside or outside the place.

By the end of the teaching, students were able to think critically about 'exogenous factors'. Students applied concepts such as racial capitalism and gentrification to explore how competing senses of place developed in both their hometown and London. Students assessed the value of concepts such as non-elite cosmopolitanism for understanding lived experiences in these places (Rogaly, 2020). They appeared to draw stronger inferences about these relationships becoming more confident in their use of geographical concepts. It was beyond the scope of the project to identify how students' capabilities developed, but there is no doubt here that the GeoCapabilties approach helped students gain access to PDK. It did so by helping the teacher think deeply and critically through collaborative engagement with geographers and other teachers. This allowed Daniel to work with the students, recontextualising geography – through the pedagogic recontextualising field – in ways that were meaningful as they supported students in thinking about their local town as 'home'.

Case study 2 (the Netherlands) – migration, feeling at home, and belonging

In Year 2 (13–14-year-olds) Anita has relative freedom to teach different topics, often guided by textbooks, Klaas for Final Year 4 (15–16-year-olds) has to teach a tight examination programme. Migration is part of the teaching in both years. The teachers saw two important challenges for their teaching. The first is how to make the abstract concepts and notions of migration meaningful for these students and their lives. Looking at a standard geography textbook in Year 1, students are asked to learn many abstract concepts in a very short time: migration, immigration, emigration, economic migration, political migration, migrants, labour migrants, regions of origin and destination, push and pull factors, and so forth. The second challenge is how to deal with the 'now and then' stereotypical textbook images of migrant neighbourhoods – places where these students live – and texts with negative connotations of unsafe and poor places with lower socioeconomic status and migration backgrounds. Behind this last challenge is the increased voice of anti-migrant standpoints in public and political debates (De Beer and De Valk, 2019) and how this affects the students at school. So, the teachers decided to connect the national curriculum to the migrant stories of the students' families or relatives in a more systematic way and to use the concept and idea of (feeling at) home.

In the vignette connected to this project, the concept of home is explored. Inspiration was found in a TED Talk of Taiye Selasi (2014):

> She states that the question 'Where are you from?' usually puts you in the box of a country, often the country on your passport, or even the country of origin of your parents, whether you come from China, the Netherlands, Morocco, Germany, Pakistan or the United States. For a lot of people, that doesn't say much about who they are. Is it clear that if you are introduced as a Dutchman or Turk, what kind of person you are? Selasi argues that for many people, we might be more of a 'local' than a 'national'. And that the places (cities, neighbourhoods) we lived in, contribute to our identity. Selasi raises the questions: where are you a local? Where is your home? To answer these questions Selasi identifies three characteristics (3 Rs) of being 'at home' or 'local': rituals, relationships and restrictions. . . . You can feel at home, feel like a local at more than one place. Because you were born in a different place or lived in different places, because you are in a different place regularly or once a year, for example. If you consider the 3 Rs for yourself, then perhaps a completely different picture will arise of where you come from, where you feel at home and who you are.
>
> *(Palings, 2021)*

The teachers designed an assignment (for both year groups) where students explored migrant stories and feelings of (being at) home through interviews. The research assignment for the students included questions about 'home', such as: what is home for them and for the persons they interviewed; what they like about home; who is important to them, if they would like to stay in that place in the future and so forth. This all refers to work on transnationalism locality, home, and identity (Blunt and Dowling, 2006).

The teachers started their course with a migrant story and selected the celebrity rapper Boef, who is very popular but also with a controversial reputation. He was born in Aubervilliers (near Paris) with a French passport from Algerian parents, moved to Eindhoven (the Netherlands) at the age of 4, lived in Houston, Texas for two years and then moved to Alkmaar (the Netherlands) at the age of 13. The picture of Boef used in the classroom shows the rapper wearing a football jersey of 'AZ' (a Dutch team from the city of Alkmaar) with 072 printed on his back, an indication for the area code of Alkmaar. The teachers raised questions about his migrant history and in Year 4 class they discussed his layered identity. In the Year 2 class, the teacher showed a short film about guest workers[6] coming to the Netherlands in the 1960s – their grandparents' generation.

The geographical concepts of home, locality, and identity were not explicitly taught. Partly because these students find it difficult to learn and use abstract concepts, and the Year 2 students were studying migration for the first time. Moreover, the teachers did not prioritise these concepts due to a lack of time. For the teachers it became clear that the students learnt a lot about their family stories and how to place these stories in the context of the wider history of labour migration to the Netherlands. But afterwards they also expressed an important shortcoming of the lessons: room for sharing, interpreting, and debriefing the findings and stories of the students in a class meeting. This is unsurprising as the work of the students offered numerous opportunities for this, showing the significance of this geography to their lives.

The students came up with very open and rich stories in their written reports. Related to Selasi (2014), the students wrote about local identification by mentioning specific neighbourhoods, their city as their home, and a wish to stay here. Relating to the 3 Rs of rituals, relationships, and restrictions (Ibid.), students considered the following question, 'when will someone feel at home in the Netherlands'? In their written assignments they mentioned rituals as 'being able to speak the Dutch language', relations with statements as 'when your family is here', and restrictions with examples as 'when you possess a passport', 'if you are treated equally', 'if you feel accepted'.

As the teachers explained, in classroom conversations, students were able to explain why they feel at home in the Netherlands and in their city. The 12–13-year-olds connected the presence of family and friends and their own house to the feeling of being at home. The 15–16-year-olds have a broader interpretation of the concept: social relationships (family and friends), their house, a safe place, a place for their activities, properties, and important for them as well is 'the place where they envision a future for themselves'. They show multi-layered identities in their stories in their written assignment.

Looking back at the course a few weeks later with two students from the Year 4 class –both with grandparents who moved from Morocco to the Netherlands –

showed the difficulty these students encounter with geographical language. The students were asked by the teacher to do a concept mapping exercise. One of the students needed a reminder of what migration is. After a brief explanation both students were able to complete the exercise, showing how they linked all the abstract concepts to the migration history of their grandparents. They even added the concepts of 'guest workers' and 'family reunification' to the map.

For most of the students, getting engaged in migration stories did contribute to the enrichment of some capabilities: for example, 'to identify choices in your life', such as to migrate themselves in future and the consequences of these choices. They learnt to change perspectives, as one of the two students in the evaluation said:

> Home is here – for us. For my parents? Home will always be where they came from. They feel at home there. It is a different kind of home. You have (a) home, but also home that really feels as home. Students were able to understand the differences of their attachment to their place here (home) and in the country of origin of their (grand)parents (holidays) and how this could differ for their parents and grandparents. They also realised that they valued things differently, such as the personal freedom they experience in the Netherlands.

Conclusion and discussion

> In his constructive-critical didactic theory, Klafki points out that the epoch typical key problem has unique educational potential for children and young people, and should be the focus of the curriculum.
>
> (Bladh et al., 2018: p. 7)

Migration can be seen as such an 'epoch typical' key problem and indeed has been used to enact a unique educational potential. Teachers and students have stressed the importance of teaching about migration. The GeoCapabilities approach as developed in Phase 3 has helped teachers to make their teaching selections about migration, powerful and meaningful. Important for this was opening up access to academic knowledge and collaborative thinking on pedagogy with geography education researchers. Moreover, the teachers used the opportunity to link their teaching to their ideas of students' needs. We see this as a form of teacher scholarship which enacts the argument made by Healy (2022) for engagement with educational theories (PDK and curriculum making in this case) combined with engagement with the discipline of geography.

At the end of the project, Daniel concluded that the GeoCapabilities approach had helped him think about the familiar topic of migration in new ways. It did so by providing a stimulus to engage with academic geography and to try out new approaches with his students, involving them in evaluating the results. Daniel's work on the project reflected and also extended his teacher scholarship. One example of this is his ongoing focus on the role of young people in the recontextualisation of knowledge (Whittall, 2021a). This leads us to reflect that Bernstein (1990) has helped draw our attention in the GeoCapabilities 3 project to the roles of teacher and student in the recontextualisation and reproduction of knowledge for opening up new ways of thinking about migration in geography.

An important element, Daniel noted, was the collaboration with other teachers and academic geographers and geography teacher educators which made the work fulfilling and helped sustain his curriculum making. In their final interview, the Dutch teachers concluded that the GeoCapabilities approach was important because it stimulated their critical reflection on what they do and why. It also increased their enthusiasm and self-confidence. They realised that to improve their lessons they needed to dedicate more time in class for reflection and interaction with the students. However, combinations of a high teaching load, students with limited language skills, with limited cultural capital, and 'schools under pressure' (Biddulph et al., 2020) make it very difficult for teachers to develop that scholarship.

The GeoCapabilities 3 project shows a form of teacher scholarship which can make geographical knowledge more accessible – and in a form that has real significance for students' capabilities (Lambert and León, 2022). However, teachers need more time to really engage with and look for new and relevant geographical knowledge to be selected and transformed into school knowledge. Investing in teachers' capacity to lead curriculum innovation is crucial to increase students' access to powerful geographical knowledge. Helping teachers form networks with each other and academic geographers and geography educators is important but so too is the school culture in which teachers work. There are control levers that can be used to support this capacity building, both at national policy level and school level. Young and Lambert (2014) envisioned a 'future school' culture which values (a progressive version of) subject knowledge. Such schools may not yet be the norm, but they are a model that is worth pursuing. We can envisage the teacher scholarship we advocate here, thriving in such schools.

Notes

1 For a full bibliography of GeoCapabilities research, see https://www.geocapabilities. org/wp-content/uploads/2021/08/GeoCap3-bibliography.pdf.
2 See Biddulph et al. (2020) https://*www.geocapabilities.org/migration-survey/* and Biddulph et al. (2020).
3 Eurogeo is the European Association of Geographers (EUROGEO), with the aim to advance the status of geography. See www.eurogeography.eu/
4 See https://www.geocapabilities.org/geocapabilities-3/
5 The teachers' own full accounts including the curriculum artefacts they developed are available as 'storymaps' at https://www.geocapabilities.org/storymaps/.
6 Guest workers, in Dutch 'gastarbeiders', are temporary labour migrants recruited in post-war Europe. In the 1970s, family reunification was allowed and many migrants stayed and became ethnic minority groups in society.

References

Béneker, T. (2018) *Powerful knowledge in Geography Education*. Utrecht: Inaugural lecture October 16th. Utrecht University

BERA (2018) *Ethical Guidelines for Educational Research*. Fourth edition.www.bera.ac.uk/resources/all-publications/resources-for-researchers (accessed 09/12/2021)

Bernstein, B. (1990) *The Structuring of Pedagogic Discourse: Vol IV, Class, Codes and Control*. London: Routledge

Biddulph, M., Béneker, T., Mitchell, D., Hanus, M., Leininger-Frézal, C., Zwartjes, L., & Donert, K. (2020) Teaching powerful geographical knowledge – A matter of social justice: Initial findings from the GeoCapabilities 3 project, *International Research in Geographical and Environmental Education*, 29(3) pp260–274.

Bladh, G., Stolare, M., & Kristiansson, M. (2018) Curriculum principles, didactic practice and social issues: Thinking through teachers' knowledge practices in collaborative work, *London Review of Education*, 16(3) pp398–413 https://doi.org/10.18546/LRE.16.3.04

Blunt, A., & Dowling, R. (2006) *Home. Series Key Ideas in Geography*. Milton Park & New York: Routledge.

Boccagni, P. (2017) *Migration and the Search for Home: Mapping Domestic Space in Migrants' Everyday Lives*. New York: Palgrave Macmillan.

Brooks, C. (2010) Developing and reflecting on subject expertise. In C. Brooks (ed), *Studying PGCE Geography at M Level: Reflection, Research and Writing for Professional Development*. Abingdon: Routledge, pp66–76.

Bustin, R. (2019) *Geography Education's Potential and the Capability Approach: GeoCapabilities and Schools*. London: Palgrave Macmillan.

Castles, S., de Haas, H., & Miller, M. J. (2020) *The Age of Migration: International Population Movements in the Modern World*. Sixth edition. Basingstoke: Palgrave Macmillan.

Catling, S., & Pike, S. (2022) Becoming acquainted: Aspects of diversity in younger children's geographies, in Hammond, L. Biddulph, M., Catling, S. and McKendrick, J. H. (eds) *Children, Education and Geography: Rethinking Intersections*. Abingdon: Routledge.

Cresswell, T. (2006) *On the Move: Mobility in the Modern Western World*. Abingdon: Taylor & Francis.

De Beer, J. A. A., & De Valk, H. A. G. (2019) De ontnuchterende rol van de wetenschap in het migratiedebat. *Demos: bulletin over bevolking en samenleving* 35(6). https://nidi.nl/demos/de-ontnuchterende-rol-van-de-wetenschap-in-het-migratiedebat

Fathi, M. (2021) Home-in-migration: Some critical reflections on temporal, spatial and sensorial perspectives. *Ethnicities* 0(0) pp1–15. http://dx.doi.org/10.1177/14687968211007462

Hammond, L. (2021) London, race and territories: Students's stories of a divided city, *London Review of Education*, 19(1) pp1–14. https://doi.org/10.14324/LRE.19.1.14

Healy, G. (2022) Geography and geography education scholarship as a mechanism for developing and sustaining mentors' and beginning teachers' subject knowledge and curriculum thinking, in Healy, G. Hammond, L. Puttick, S. Walshe, N. (eds) *Mentoring Geography Teachers in the Secondary School: A Practical Guide*. Abingdon: Routledge

Jazeel, T. (2019) *Postcolonialism*. Routledge: London and New York.

Klafki, W. (2000) Didaktik analysis as the core of preparation of instruction, in Westbury, I., Hopmann, S. and Riquarts, K. (eds) *Teaching as a Reflective Practice: The German Didaktik tradition*. London: Routledge, pp139–160.

Lambert, D., & León, K. (2022) The value of geography to on individual's education, in, Hammond, L. Biddulph, M., Catling, S. and McKendrick, J. H. (eds) *Children, Education and Geography: Rethinking Intersections*. Abingdon: Routledge.

Lambert, D, Solem, M., & Tani, S. (2015) Achieving human potential through geography education: A capabilities approach to curriculum-making in schools, *Annals of the Association of American Geographers*, 105(4) pp723–735.

Massey, D. (1994) *Space, Place and Gender*. Cambridge: Polity Press.

Maude, A. (2016) What might powerful geographical knowledge look like?, *Geography* 101(2) pp70–76.

Mitchell, D. (2020) *Hyper-socialised – Enacting the Geography Curriculum in Late Capitalism*. Abingdon: Routledge.

Nieto, M. (2021) Temporalities. In: Nieto M., Massa A., & Bonfanti S. (2021) *Shifting Roofs: Ethnographies of Home and Mobility*. Oxon: Routledge. pp91–114.

Palings, J. (2021) Where are you from or where are you at home? www.geocapabilities.org/vignettes/where-are-you-from-or-where-are-you-at-home/

Rogaly, B. (2020) *Stories from a Migrant City Living and Working Together in the Shadow of Brexit*. Manchester: Manchester University Press

Schapendonk, J. (2020) *Finding Ways Through Eurospace*. West African Movers Re-viewing Europe from the Inside. Oxford: Berghahn Books

Selasi, T. (2014) Transcript Tedtalk Don't ask me where I'm from. Ask me where I'm local. TED-talk. www.ted.com/talks/taiye_selasi_don_t_ask_where_i_m_from_ask_where_i_m_a_local/discussion [Accessed 17 March 2022].

Van Liempt, I., & Staring, R. (2021) Homemaking and places of restoration: Belonging within and beyond places assigned to syrian refugees in the Netherlands. *Geographical Review*, 111(2) p308–326. https://doi.org/10.1080/00167428.2020.1827935

Walker, M. (2006) Towards a capability-based theory of social justice for education policy making, *Journal of Education Policy*, 21(2) pp163–185

Whittall, D. (2021a) The role of students in the recontextualisation and transformation of powerful knowledge: A study of sixth form geography students, In Fargher, M., Mitchell, D., & Till, E. (eds) *Recontextualising Geography in Education*. London: Springer Nature.

Whittall, D. (2021b) *A critical understanding of place*. www.geocapabilities.org/vignettes/critical-understanding-of-place-migration/

Wilkins, A. (2019) *Migration, Work and Home-Making in the City: Dwelling and Belonging among Vietnamese Communities in East London*. Abingdon: Routledge.

Young, M., & Lambert, D. (2014) *Knowledge and the Future School – Curriculum and Social Justice*. London: Bloomsbury.

13

LOOKING CLOSELY FOR ENVIRONMENTAL LEARNING

Citizen science and environmental sustainability education

Ria Dunkley

Introduction

Within this chapter, I will determine the effects of participating in environmental citizen science as a child, specifically effects related to environmental learning. I will discuss how environmental citizen science can be used by teachers of geography and geography teacher educators within formal and informal educational settings, including schools, nurseries, children and youth groups, and play-based learning settings. I will rethink the intersections between informal and formal learning environments and informal and formal pedagogies by exploring these potentials (Biddulph et al., 2022). The chapter contributes to understanding the intersections between children, geography, and education, specifically through understanding how learning within the time and space of a child's everyday life interacts with learning within a more formal setting. Moreover, the chapter highlights how teachers of geography can shape geography education by understanding how children's and young people's geographies can be enhanced through citizen science participation and how to harness its potential for environmental sustainability education. Specifically, I will discuss how micro-ecological worlds and the 'micro-geographies' (Kraftl, 2022) of the school ground and local green, blue, and grey spaces can become the focus of citizen science projects that utilise technology. Citizen science is a practice that can enhance human appreciation of our intertwined relationship with non-humans through making invisible aspects of micro-worlds and non-habitats visible to the children who become involved in citizen science projects.

This chapter will explore the potential for environmental learning within an environmental citizen science project that collects data on foraging sources for bees in urban environments. Drawing upon an example of an ecological citizen science project, 'Spot-a-Bee', I will discuss how citizen science might play a vital role in

DOI: 10.4324/9781003248538-17

making nature connections visible to young people. I will pay particular attention to environmental learning, human/non-human relationality, and the cognisant benefits of sustainable behaviours for individuals and communities. I argue that guiding children to notice, name, record, and reflect upon the presence (or absence) of bee-friendly habitats within urban contexts can increase ecological consciousness. Furthermore, that plays a part in nurturing values and attitudes that are more harmonious within more-than-human worlds.

This chapter will describe the four-stage process of Spot-a-Bee and show how the embodied monitoring processes involved in citizen science can make familiar environments encountered everyday unfamiliar to participants. Through a four-stage process of observing, capturing, recording, and reflecting, I argue that environmental citizen sciences like Spot-a-Bee enable intimate encounters with flowers and bees, which raise the consciousness of the non-human species with whom we share our specific places. Then, through sharing photographs with the Spot-a-Bee community through the Spot-a-Bee App and web-map, a consciousness of how the bee-places to which we are intimately connected compare with other bee-places across the UK. Finally, there are opportunities to reflect upon our bee-places in the wider UK and global contexts as the data bank grows to consider the coupled climate and ecological emergency.

In revealing such insights, within this chapter, I question the assumption that citizen science is an activity that mainly benefits the scientists behind the project and leads to predetermined outcomes for participants. Instead, I reveal an emergent, place-based, and highly personalised citizen science experience for children who participate. In doing so, I highlight the underexplored benefits of citizen science in climate change education and ecopedagogy (Dunkley, 2018).

What is environmental citizen science?

Environmental citizen science is a practice of open science that involves interested people in collecting environmental research. The process of participating in environmental citizen science is valuable, in and of itself, as a process that focuses human attention on the natural environment. It is also a collaborative activity (Dunkley, 2019) that enhances a person's social connections (Price and Lee, 2013) to those running projects and the wider project community of contributors. The broader effects of participation in activities that enhance place and social connections are thought to enhance human connections to the environment for positive mental health and well-being (Dunkley, 2019; Geoghegan et al., 2016). Citizen science also enhances environmental learning (Dunkley, 2017; Miller-Rushing et al., 2012; Parliamentary Office of Science and Technology, 2014) and can nurture pro-environmental behaviours (Dickinson et al., 2013). Nevertheless, the potential of citizen science as an activity that enhances the ecological awareness of young people is under-explored in the academic literature. Furthermore, studies that focus on the benefits of citizen science for individuals and communities, barring a few examples (Busch, 2013; Dunkley, 2018), are also lacking (Geoghegan et al., 2016).

The digital revolution spurs citizen science at a scale and scope previously unimaginable. During the COVID-19 lockdowns, citizen science projects have grown in popularity within the UK, leading to local benefits such as burgeoning awareness and the use of urban ecologies. Nevertheless, a place-based lens allows us to appreciate what citizen science does for people and non-humans in local contexts. In a recent study (Dunkley, 2019), I explored volunteer involvement in environmental citizen science initiatives. In doing so, I understood the socio-geographical influences that act on participation in environmental citizen science.

Furthermore, I proposed that affective connections with local geographies provided a conceptual framework for understanding citizen science motivations. This provision study revealed that early affective bonds formed with ecological spaces endured throughout life courses while citizen science participation offered a way of remaining connected to local environments. It is a process that enables participants to observe ecological surroundings, participate in environmental research, and protect local environments. Citizen science emerged from this previous study as a framework to connect to and protect local and global affinity spaces while monitoring global environmental change. This process is summarised in Figure 13.1.

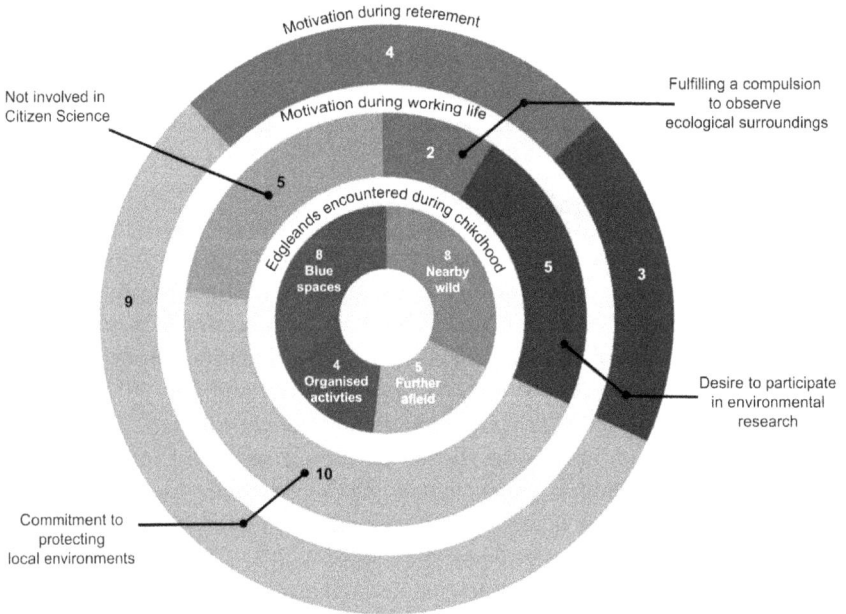

FIGURE 13.1 Spaces encountered during childhood and citizen science motivations in working life and retirement (counts are based on references within all 22 narratives and are not mutually exclusive)

Source: extracted from Dunkley, (2019).

Twenty-first-century childhoods, climate and ecological emergency, and citizen science

For the adults interviewed as part of this study, many of whom were in retirement, the opportunity to participate in organised citizen science had not been an option during their childhood. Today, opportunities to participate in citizen science have been dramatically increased through globalising influences, in particular, the digital revolution of the 1980s and the advent of the internet. Over 3,000 citizen science projects are listed globally on the citizen science portal 'SciStarter' (https://scistarter.org). As a result of living in an increasingly digital and globally connected society, children in the twenty-first century can contribute to place-based environmental citizen science projects on a wide range of topics – from bee and butterfly monitoring to forest and river monitoring.

At the same time, being a child in the present moment means living with a great deal of uncertainty in the face of an environmental and climate emergency. Many communities are directly feeling the effects of these coupled emergencies, as they are playing out on screens for those who are one step removed. Many communities globally are the direct impacts of climate change, for example, flooding of their homes and community spaces, resulting from increased sea level rise. Simultaneously, climate and eco-anxiety are on the rise amongst younger generations, many of whom experience a sense of helplessness in the face of inaction on the climate crisis (Randall and Brown, 2016). Citizen science was a means by which adults in my 2018 study came to connect to and protect local and global affinity spaces while assisting in monitoring global environmental change (Dunkley, 2018). Also, it can offer children living with climate and ecological emergencies the same unique affordances.

Hope and citizen science

Environmental citizen science involves direct observation, capturing, recording, and reflecting on the natural world. The benefits of spending time establishing nature connections are widely described (Holland et al., 2021; McCrorie et al., 2021; Louv, 2008; Richardson and Sheffield, 2017). Nevertheless, some benefits come from capturing, recording, and reflecting upon the ecologies surrounding us. This practice strengthens and deepens our knowledge, understanding, and confidence in interacting with the non-human world. Therefore, at an initial level, citizen science may offer an opportunity to 'make kin' (Haraway, 2016) with the non-human world, that is, to get to know our non-human neighbours better. Haraway (2016: p. 12–13) has stressed how, during times of ecological crisis especially, it is essential to recognise our intertwined existence as companion species, humans, and non-humans, with whom we 'are relentlessly becoming with'.

Nevertheless, what is perhaps explored to a lesser extent is an awareness of the effects of data generated through citizen science for those participating. Citizen science data can influence planning decisions at a local level. For example, Newman et al. (2017) discovered that citizen science projects that leveraged place-based networks and place dimensions were utilised in conservation decision-making. Citizen science data can potentially contribute to understanding human impact upon biodiversity and climate change on a global scale. For example, it could contribute valuable understandings to the International Panel of Biodiversity and Ecosystem Services' global data on the state of biodiversity and the ecosystems services provided to society (Dunkley et al., 2018). Contributing to local and global-level ecological understandings is potentially empowering for those who participate. Participants' observations may potentially play a part in protecting the local spaces they are connected to and influencing understanding and action upon the local and global issues about which they are often passionate. The benefits that citizen science affords are additional to enhancing nature connectedness and may relate to generating nature confidence and enhancing a sense of agency (Dunkley, 2019).

However, citizen science may also have another significant role in times of climate and ecological emergency. Haraway (2016: p. 1) urges us to develop lines of 'inventive connection as a practice of learning to live and die well with each other in a thick present'. She states that, in the context of climate and ecological emergency: 'our task is to make trouble, to stir up potent response to devastating events, as well as to settle troubled waters and rebuild quiet places'. In enacting citizen science, individuals regard citizen science as a 'response-ability' (Haraway, 2016) to environmental injustices through the generation of collective knowledge that will facilitate conservation science. In this sense, citizen science is a hopeful practice rooted in local action. It is a way of being aware of ecological emergencies within a local context while purposefully capturing, recording, reporting, and reflecting on what is found.

Citizen science as an (eco)pedagogy of hope?

I am turning now to consider childhood involvement in citizen science. Citizen science can be considered a part of ecopedagogic praxis (Kahn, 2008; Misiaszek, 2019). Ecopedagogues seek to raise ecological consciousness. Raising ecological consciousness is conceptualised here as a process the aim of which is to recognise the immersive relationship that humans have with nature. Through raising awareness of human–nature relationality, ecopedagogy highlights both the anthropogenic basis of the coupled climate and ecological emergency and the self-destructive nature of these coupled emergencies (Dunkley, 2018). It seeks to inspire and facilitate transformative local and global action on ecological crises and climate emergencies while also reflecting upon the systemic underpinnings of these crises (Misiaszek, 2015). It builds upon Freire's (2014) critique of the 'banking approach' education, which regards children as passive recipients of adult/teacher knowledge. Instead, ecopedagogy recognises the grounded indigenous and place-based knowledge systems within the specific learning context encountered. Ecopedagogy also seeks to critique dominant knowledge, which may have characterised children's education to date and the effects of such knowledge in driving, facilitating, or justifying environmental and social crises in direct and indirect ways (Misiaszek, 2016; Omiyefa et al., 2015). Such dominant knowledge is informed by cultural ideologies and values emergent from the Global North, which may reflect a paradigm that reinforces the status quo, in terms of human–non-human relationality and marginalise certain groups within society. Critical pedagogy's role in transforming the lived experiences of individuals in society makes ecopedagogy, as its parent concept is a pedagogy of hope (Freire, 2014).

Citizen science is an opportunity to enact ecopedagogy through its dual capacity that it offers children to take an ecocentric lens, which presents an opportunity to gain an appreciation of their relationship within the non-human world (Barratt Hacking and Taylor, 2020; Cutter-Mackenzie-Knowles et al., 2020). Involving children in citizen science could offer the opportunity for children to engage in the form of non-violent direct climate and environmental action (Butler, 2021). Eco-anxiety is thought to be a sense of powerlessness about what to do about climate change (Randall and Brown, 2016), coupled with a sense that others are not willing to act (GAP, 2021). Eco-anxiety is thought to affect a growing number of children and young people. The opportunity within citizen science participation to actively participate in understanding local environments as part of broader conservation activities could have important implications for children's mental health.

Citizen science is also a practice rooted in resisting the grand narratives about children needing to be educated for sustainable development given their status as symbolic members of an imagined collective future. Resisting such grand narratives is crucial in times of climate and ecological emergency. Emphasis is often placed in many education for sustainability initiatives upon sustainable futures and future generations. Emphasising future actors is problematic given that evidence shows that 'temporal psychological distancing' (Spence, Poortinga and Pidgeon, 2012) of climate and ecological crises increases the sense of powerlessness that individuals feel in the present. Citizen science is a practice rooted in present-day actions to address environmental issues at a local and global scale. As a result, involving children in citizen science could help limit the prevalence and effects of eco-anxiety amongst young learners through its capacity to focus a child's attention on how scientific knowledge-building plays a prominent role in addressing the causes and effects of climate change and ecological issues.

Moreover, through making science accessible, citizen science can be thought of as forming part of a pedagogy of hope (Freire, 2014) – expanding contributions in scale and scope, thereby considering structural inequalities that exist in terms of the generation of scientific knowledge. Citizen science, often undertaken via digital platforms, can be a democratising force. Schools now emphasise digital literacy, and many classrooms will have access to tablets, laptops, and other digital devices. However, offline citizen science is also possible, given that many citizen science projects collect data using paper-based approaches or offer a hybrid approach, which provides access to online and offline data collection tools. The democratising ethos of citizen science can be brought to the classroom by teachers of geography. It can, as a process, challenge the elitist nature of traditional scientific endeavours and open up science beyond gender and race boundaries.

A working example of citizen science: Spot-a-Bee

To give a specific example of how citizen science can enhance the teaching of geography, I will now introduce Spot-a-Bee, a citizen science project established in collaboration between life and social scientists in 2016. The purpose of the Spot-a-Bee App is to collect and visualise environmental observations through photographs. Users also provide geographical references and written descriptions of their observation, including plant and bee species, weather conditions, the date, and the time. Since launching the App in May 2020, it has received over 11,000 contributions. The context in which Spot-a-Bee was developed is increasing biodiversity loss, driven by anthropogenic climate change and urbanisation processes (Sanchez-Ortiz et al., 2019). Many UK bee species are in decline (Powney et al., 2019), partly due to the loss of flower-rich habitats and plant diversity (Vanbergen and Initiative, 2013). Urban ecologies are increasingly recognised as providing essential habitats for bees while most human–nature encounters, including encounters with bees and the plants they rely upon, occur within urban contexts (Kuras et al., 2020).

At the same time, mass species extinction is increasingly linked to human estrangement from nature (Haraway, 2016; Latour, 2011; Clarke and Witt, 2022). It has been suggested that increasing human–nature encounters enhance nature connectedness (Martin et al., 2020), which increases the likelihood of pro-environmental behaviours and actions that could positively impact biodiversity (Haraway, 2016; Kohn, 2013; Latour, 2018; Tsing, 2015). Connecting to local urban ecologies is beneficial for mental health and well-being (Sandifer et al., 2015) and environmental learning (Sageidet et al., 2018). In the context of this understanding, driven through interdisciplinary collaboration, Spot-a-Bee continues to grow.

The pedagogy of Spot-a-Bee

Spot-a-Bee has four stages, as represented in Figure 13.2, and each of these stages leads to specific pedagogic benefits, as will be described in this following section.

Stage one – observe

Social scientists and public health researchers have identified that greater access to green space is beneficial for psychological restoration. Studies have also evidenced the mental health, well-being, and environmental learning benefits of looking at plants (Haviland-Jones et al., 2005), often overlooked (Parsley, 2020; Wandersee and Schussler, 1999). Downloading the Spot-a-Bee App is an invitation to follow bees to identify the plants they rely upon in urban environments. It is also a way of 'making kin' (Haraway, 2016) amongst human and non-human neighbours within our neighbourhoods. It is a way of making connections meaningful in a broader context. There are multiple benefits of this stage in the Spot-a-Bee process. It has been argued that in Western societies, we ignore plants, which has resulted in

Observe

Reflect   Capture

Record

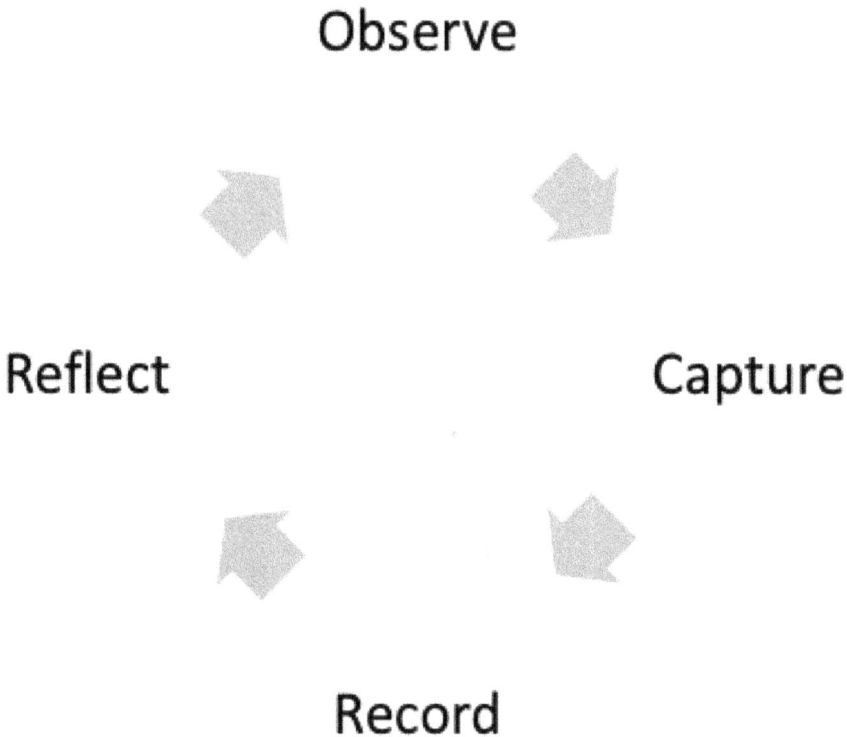

FIGURE 13.2   The four stages of Spot-a-Bee

'plant blindness' (Wandersee and Schussler, 1999) or 'Plant Awareness Disparity' (Parsley, 2020). While participating in Spot-a-Bee, bees guide children's attention towards the urban landscape in gardens and allotments. In doing so, children have the opportunity to experientially encounter the scientific knowledge that such environments are known to support greater bee species than non-urban environments (Baldock et al., 2019). In urban schools, Spot-a-Bee may also guide children's attention towards the plantscapes that might be growing out of the sides of school buildings, hidden in railway sidings or upon urban wastelands near schools.

Step 2: capture (photography skills; digital literacy)

As Ouvry (2020, p. 81) states, successful outdoor learning depends on 'the capacity of what is there to be noticed by the children (and adults), to capture their imagination and curiosity so they want to look further'. The process of capturing an observation is critical to the citizen science process, both for scientific outcomes and participant learning. This process of capturing is one that, within Spot-a-Bee, is made possible, primarily through a mobile phone. It is increasingly acknowledged in many regions of the world. Digital devices are a ubiquitous part of the modern lives of children and adults alike. Contrary to narratives that position digital devices as a diminutive of children's time outdoors, they can result in more attention being paid to spaces, places, and more-than-human worlds. Indeed, more nuanced understandings are now emerging of how digital devices augment the experience of being outdoors (Nielsen et al., 2021; Smith and Dunkley, 2018).

The process of capturing bee images presents additional benefits of citizen science participation, including skills enhancement (POST, 2014) and enhancement of digital literacy and knowledge development (Dickinson et al., 2012). The specific skills that children stand to gain and enhance through participating in Spot-a-Bee include nature observation, including the ability to identify and categorise flora and fauna. Gaining these specific skills is significant given the lack of natural history within formal education in the UK, which has led to recent calls for a GCSE in natural history to be created (Moore-Anderson, 2021). Furthermore, children's confidence in the practice of digital photography may be enhanced.

Step 3: record

In terms of environmental learning and literacy, spotting activities provide participants with a valuable opportunity to give their attention to plants. Recording the existence of a plant in a local area can be considered a step towards forming an alternative relationship with plants. It could encourage a deeper relationality with the non-human world, which counters the existing experiences of many young people growing up in Western cultures. At one level, participating in Spot-a-Bee contributes to acquiring diminished 'plant literacy' (Parsley, 2020; Wandersee and Schussler, 1999). When a participant has taken a photograph, they use a drop-down list in the App to identify both the plant and the bee (see Figure 13.3). If it is an unusual bee or plant, this may require further taxonomic research, through which there are opportunities to enhance the ecological learning process.

The recording stage of Spot-a-Bee involves using mobile Apps to understand pollinators' presence (and absence) in local contexts. Doing so can deepen an appreciation of the coexistence of species within the place. Kimmerer (2020: p. 216) writes that field trips to nature reserves and national parks have a role in enhancing learners' sense of 'botanical belonging' (Kimmerer, 2020: p. 216). She reflects that species classification was only one aspect of the experiential learning that happens during field trips. The deeper affective dimension of the learning experience was that it enabled learners to give their attention to other species and that paying attention was 'a form of reciprocity with the living world' (Kimmerer, 2020: p. 222).

Similarly, citizen science projects, such as Spot-a-Bee, can be thought of as providing children with the opportunity to encounter the non-human world and give their attention to it. By paying attention to the broader urban ecosystem within which they coexist with other species, participants begin to know their local environments differently. Within this context, App-technology has the potential to

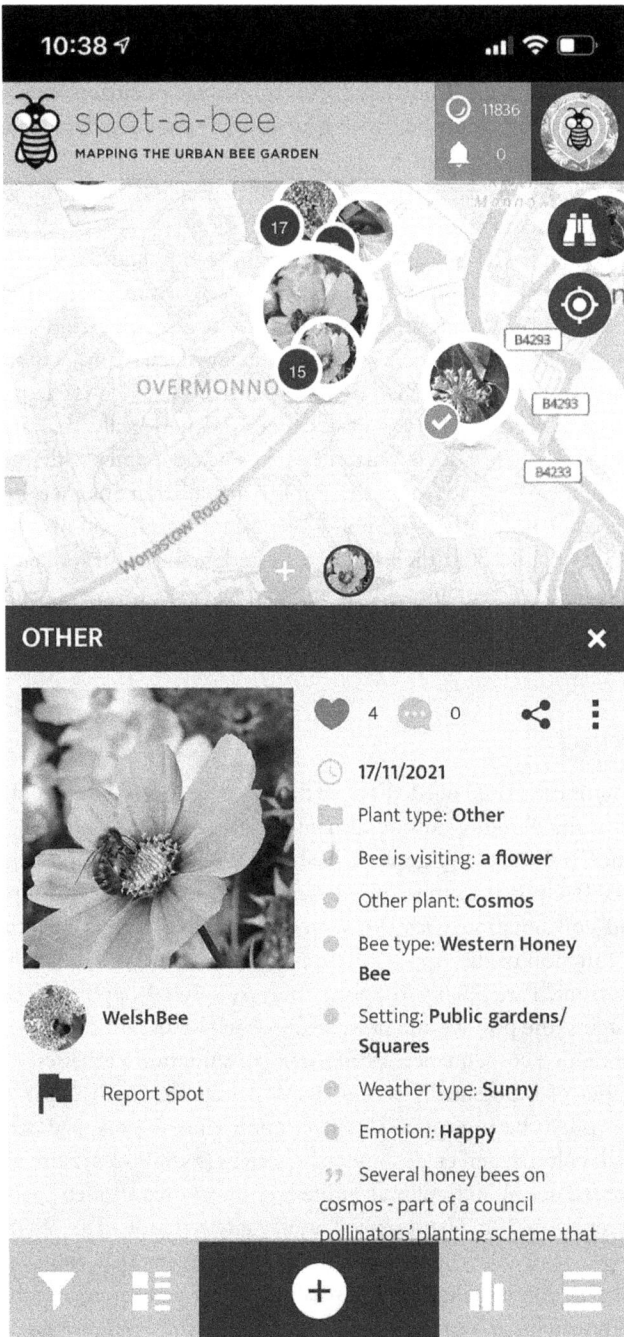

FIGURE 13.3   Spot-a–Bee App record

extend such learning, allowing children and young people to appreciate, through enablers including geotagging, broader intertwined biodiversity contexts in real time, in a way that classroom-based science education fails to achieve (Rousell and Cutter-Mackenzie-Knowles, 2020). In the case of citizen science, access to technology enhances relationships with nature rather than detracting from them, as people like Louv (2012) propose.

In primary and secondary schools, links can be made between subjects such as science, geography, and information communication technology by focussing on students' health and well-being, responsible citizenship, and active participation. The recording stage of the process can also be an opportunity for transdisciplinary learning. It can enable the provision of Climate Change Education and of Learning for Sustainability within the classroom. Citizen science projects are useful to teachers wishing to engage in these fields, given that evidence suggests that learning within them needs to be action-oriented and personally relevant (Monroe et al., 2019). There are also opportunities to engage pupils with more critical learning through citizen science. For example, participation enhances scientific literacy through the contribution of Spot-a-Bee data in graphs and heat maps. These graphs and heat maps could then be used in geography to introduce progressive ideas such as critical cartography, which attends to different ways of mapping space, in this case, according to what spaces are 'on the map' for non-human inhabitants (Haraway, 2016).

Stage 4: reflect

The most significant shift needed for societies to address climate and ecological emergencies is for thinking outside of confined disciplines (Schneider, 1977), and citizen science is particularly good at enabling this. The final and essential stage in the Spot-a-Bee process is that of reflecting. This reflection can happen in local contexts and collaboration with the wider Spot-a-Bee community through the 'comments' function of the App – where participants tend to talk to others about their observations. Participants in Spot-a-Bee are offered opportunities to reflect upon ecosystems, the positive and negative feedback loops embedded within them, and the unintended consequences of modern consumption processes.

An example of a potentially productive learning area with upper-secondary-level students might be to explore to reflect upon what the mapped data means in broader social, cultural, and environmental contexts (Figure 13.4). Urban ecologists have highlighted social and cultural influences upon urban green space management (Kuras et al., 2020). The so-called luxury effect (Hope et al., 2008) accounts for higher species diversity in affluent urban areas, given that exotic plants tend to be planted in gardens in affluent areas and more significant numbers. However, recent research has also shown that higher income levels are not necessarily positively correlated with greater species diversity (Chamberlain et al., 2019), and non-native plants may negatively impact biodiversity, given that many local insect species are specialised in native plants (Mata et al., 2021).

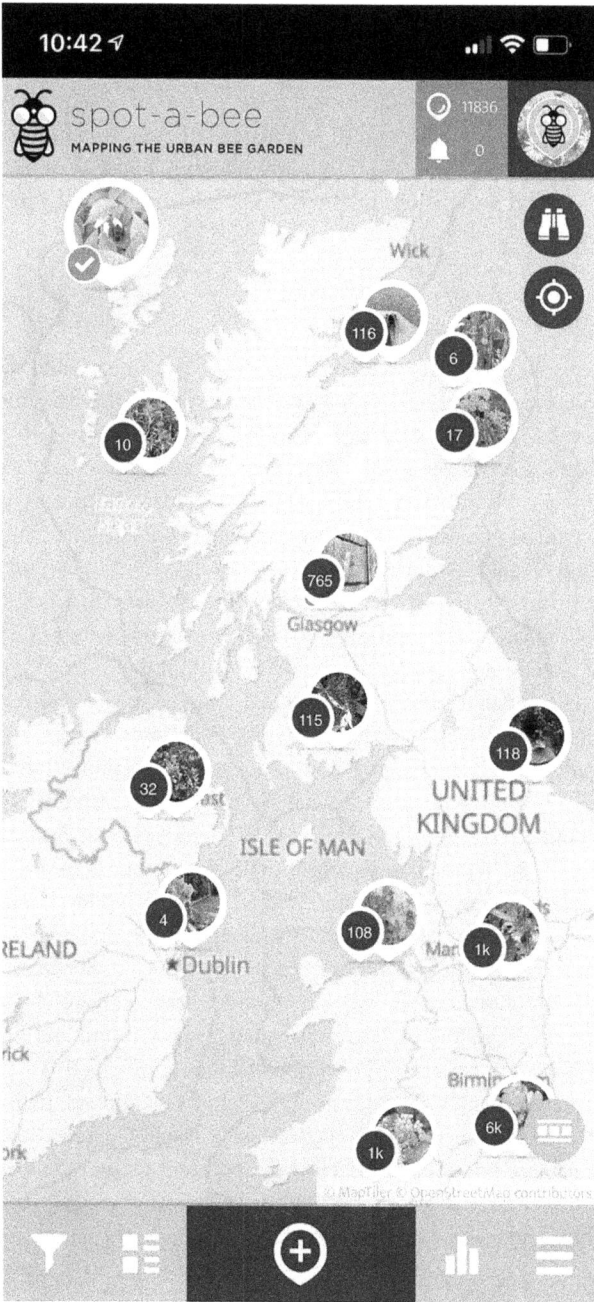

FIGURE 13.4   Spot-a-Bee UK map

Source: Spot-a-Bee App.

Then, finally, there is the opportunity for young people to achieve through the citizen science community of practice or local and global action on biodiversity loss. These have the benefits of enhancing senses of self-efficacy and improving social capital (Dickinson et al., 2013) and enhancing place connections (Dunkley, 2018). The benefits of involvement in citizen science projects can be enhanced by developing learning resources that augment citizen science. Within Spot-a-Bee, we have begun working on resources, including for families and youth groups to serve this purpose.

Conclusions

I have set out to show benefits of projects such as Spot-a-Bee beyond those preconceived by project leaders and also to highlight the always-in-the-becoming nature of educational activities. Moreover, I have provided insight into how real-world, experiential, problem-based citizen science can be integrated into the teaching of geography in primary and secondary classrooms. However, citizen science projects, such as Spot-a-Bee, could also be used by those working with children and young people in a range of settings. These may include early years settings, child and youth groups, and play-based learning settings. In such settings, educators and youth workers face the challenge of identifying place- and context-relevant citizen science projects. They are then free to augment the citizen science experience to fit their specific setting aims.

Citizen science offers children and teachers a tangible way to enhance transdisciplinary learning by employing a scientific method and mindset to observe the natural world, close and careful observation, slow action, and reflection. Project results can be placed into a broader context that enhances appreciation of human interconnections within connected ecological–social–technological systems. Furthermore, citizen science makes visible the consequences of our interventions in place – intended and unintended – through the connection between global and local context, microspecies and mega problems like biodiversity loss and climate change.

I have moreover highlighted how participating in citizen science may give people hope in local contexts while living in a world where climate change and biodiversity loss can be overawing. I have provided a sense of building knowledge in a crisis, which constitutes an ecopedagogy of hope (Freire, 2014) in the climate emergency. Citizen science offers a means for teachers of geography and a range of other subjects to enhance their ecopedagogic offering. Citizen science could be included to a greater extent in Initial Teacher Education programmes, and funding, support, and space could be given to existing teachers of geography and schools to integrate citizen science in their current practice.

References

Baldock, K., Goddard, M. A., Hicks, D. M., Kunin, W. E., Mitschunas, N., Morse, H., Osgathorpe, L. M., Potts, S. G., Robertson, K. M., Scott, A. V., Staniczenko, P., Stone, G. N., Vaughan, I. P., & Memmott, J. (2019) A systems approach reveals urban pollinator hotspots and conservation opportunities. *Nature Ecology & Evolution*, 3(3) pp363–373. https://doi.org/10.1038/s41559-018-0769-y

Barratt Hacking, E. and Taylor, C.A. (2020) Reconceptualising international-mindedness in and for a posthuman world. *International Journal of Development Education and Global Learning*, 12(2), pp133–151.

Biddulph, M. Hopkins, P. and Tate, S. (2022) Connecting children's and young people's geographies and geography education: Why this matters to and for children, education and society, in Hammond, L. Biddulph, M., Catling, S. and McKendrick, J. H. (eds) *Children, Education and Geography – Rethinking Intersections*. Abingdon: Routledge

Busch, A. (2013) *The Incidental Steward: Reflections on Citizen Science*. New Haven, CT: Yale University Press.

Butler, J. (2021) *The Force of Nonviolence: An Ethical-political Bind*. London: Verso.

Chamberlain, D. E., Henry, D. A., Reynolds, C., Caprio, E. and Amar, A. (2019) The relationship between wealth and biodiversity: A test of the luxury effect on bird species richness in the developing world. *Global change biology*, 25(9) pp3045–3055.

Clarke, H. and Witt, S. (2022) Field-visiting: paying attention with more than human worlds. In L. Hammond, M. Biddulph, S. Catling and J.H. McKendrick (eds) Children, *Education and Geography – Rethinking Intersections*. Abingdon: Routledge

Cutter-Mackenzie-Knowles, A., Malone, K. and Barratt Hacking, E. (2020), *The Research Handbook on Childhoodnature*. Cham: Springer AG.

Dickinson, J. L., Crain, R. L., Reeve, H. K. and Schuldt, J. P. (2013) Can evolutionarily design of social networks make it easier to be 'green'? *Trends in Ecology & Evolution*, 28(9) pp561–569.

Dickinson, J.L., Shirk, J., Bonter, D., Bonney, R., Crain, R.L., Martin, J., Phillips, T. and Purcell, K. (2012) The current state of citizen science as a tool for ecological research and public engagement. *Frontiers in Ecology and the Environment*, 10(6) pp291–297.

Dunkley, R., Baker, S., Constant, N. and Sanderson-Bellamy, A. (2018) Enabling the IPBES conceptual framework to work across knowledge boundaries. *International Environmental Agreements*, 18, pp779–799. https://doi.org/10.1007/s10784-018-9415-z

Dunkley, R.A. (2017) The role of citizen science in environmental education: A critical exploration of the environmental citizen science experience. In L. Ceccaroni & J. Piera (Eds.), *Analysing the Role of Citizen Science in Modern Research* (pp213–230). IGI Global. http://dx.doi.org/10.4018/978-1-5225-0962-2.ch010

Dunkley, R.A. (2018) Space-timeScapes as ecopedagogy. *The Journal of Environmental Education*, 49(2) pp117–129.

Dunkley, R.A. (2019) Monitoring ecological change in UK woodlands and rivers: An exploration of the relational geographies of citizen science. *Transactions of the Institute of British Geographers*, 44(1) pp16–31.

Freire, P. (2014) *Pedagogy of Hope: Reliving Pedagogy of the Oppressed*. London: Bloomsbury Publishing.

Geoghegan, H., Dyke, A., Pateman, R., West, S., & Everett, G. (2016). Understanding motivations for citizen science (Final report on behalf of UKEOF). University of Reading Stockholm Environment Institute (University of York) and the University of the West of England. www.ukeof.org.uk/resources/citizen-science-resources/citizen-scienceSUMMARYReportFINAL19052.pdf

Global Action Plan (2021) *Generation Action White Paper*. www.globalactionplan.org.uk/files/generation_action_white_paper.pdf, Accessed: 29/10/2021

Haraway, D. J. (2016) *Staying with the Trouble: Making kin in the Chthulucene*. Duke, NC: Duke University Press.

Haviland-Jones, J., Rosario, H.H., Wilson, P. and McGuire, T.R. (2005) An environmental approach to positive emotion: Flowers. *Evolutionary Psychology*, 3(1), p. 147470490500300109.

Holland, I., DeVille, N.V., Browning, M.H.E.M., Buehler, R.M., Hart, J.E., Hipp, J.A., Mitchell, R., Rakow, D.A., Schiff, J.E., White, M.P., Yin, J. and James, P. (2021) Measuring nature contact: A narrative review. *International Journal of Environmental Research and Public Health*, 18(8), 4092. (DOI: 10.3390/ijerph18084092) (PMID:33924490) (PMCID:PMC8069863)

Hope, D., Gries, C., Zhu, W., Fagan, W.F., Redman, C.L., Grimm, N.B., Nelson, A.L., Martin, C. and Kinzig, A. (2008). Socioeconomics drive urban plant diversity. In *Urban Ecology* (pp. 339–347). Boston, MA: Springer.

Kahn, R. (2008) 'From education for sustainable development to eco-pedagogy: Sustaining capitalism or sustaining life?' in *The Journal of Eco-pedagogy*, 4(1) pp1–8.

Kimmerer, R.W. (2020) *Braiding Sweetgrass: Indigenous Wisdom, Scientific Knowledge and the Teachings of Plants*. London: Penguin.

Kohn, E. (2013) *How Forests Think: Toward an Anthropology beyond the Human*. Los Angeles, CA: University of California Press.

Kraftl, P. (2022) 'Geographies of Education Spaces', in Hammond, L., Biddulph, M., Catling, S. and McKendrick, J. H. (eds) *Children, Education and Geography – Rethinking Intersections*. Abingdon: Routledge.

Kuras, E.R., Warren, P.S., Zinda, J.A., Aronson, M.F., Cilliers, S., Goddard, M.A., Nilon, C.H. and Winkler, R. (2020) Urban socioeconomic inequality and biodiversity often converge, but not always: A global meta-analysis. *Landscape and Urban Planning*, *198*, p. 103799.

Latour, B. (2011) Waiting for Gaia. Composing the common world through arts and politics. Lecture presented at the Launch of Science Po Programme in Arts and Politics, French Institute, London.

Latour, B. (2018) *Down to Earth: Politics in the New Climatic Regime*. Hoboken, NJ: John Wiley & Sons.

Louv, R. (2008) *Last Child in the Woods: Saving Our Children from Nature-Deficit Disorder*. Chapel Hill, NC: Algonquin Books.

Louv, R. (2012) *The Nature Principle: Reconnecting with Life in a Virtual Age*. Chapel Hill, NC: Algonquin Books.

Martin, L., White, M.P., Hunt, A., Richardson, M., Pahl, S. and Burt, J. (2020) Nature contact, nature connectedness and associations with health, wellbeing, and pro-environmental behaviours. *Journal of Environmental Psychology*, 68, p. 101389.

Mata, L., Andersen, A.N., Morán-Ordóñez, A., Hahs, A.K., Backstrom, A., Ives, C.D., Bickel, D., Duncan, D., Palma, E., Thomas, F. and Cranney, K. (2021) Indigenous plants promote insect biodiversity in urban greenspaces. *Ecological Applications*, p.e02309.

McCrorie, P., Olsen, J. R., Caryl, F. M., Nicholls, N. and Mitchell, R. (2021) Neighbourhood natural space and the narrowing of socioeconomic inequality in children's social, emotional, and behavioural wellbeing. *Wellbeing, Space and Society*, 2, p. 100051. (Doi: 10.1016/j.wss.2021.100051)

Miller-Rushing, A., Primack, R., & Bonney, R. (2012) The history of public participation in ecological research. *Frontiers in Ecology and the Environment*, 10(6) pp285–290.

Misiaszek, G.W. (2015) Ecopedagogy and Citizenship in the Age of Globalisation: connections between environmental and global citizenship education to save the planet. *European Journal of Education*, 50(3) pp280–292. https://onlinelibrary-wiley-com.ezproxy.lib.gla.ac.uk/doi/full/10.1111/ejed.12138 (Accessed 26/03/21).

Misiaszek, G.W. (2016) 'Eco-pedagogy as an element of citizenship education: The dialectic of global/local spheres of citizenship and critical environmental pedagogies'. *International Review of Education*, 62(5) pp587–607.

Misiaszek, G.W. (2019) 'Eco-pedagogy: Teaching critical literacies of 'development', 'sustainability', and 'sustainable development''. *Teaching in Higher Education*, 25(5) pp615–632.

Monroe, M.C., Plate, R.R., Oxarart, A., Bowers, A. and Chaves, W.A. (2019) Identifying effective climate change education strategies: A systematic review of the research. *Environmental Education Research*, 25(6), pp791–812, DOI: 10.1080/13504622.2017.13 60842.

Moore-Anderson, C. (2021) Putting nature back into secondary biology education: A framework for integration, *Journal of Biological Education*. https://doi.org/10.1080/0 0219266.2021.1979628

Newman, G., Chandler, M., Clyde, M., McGreavy, B., Haklay, M., Ballard, H., Gray, S., Scarpino, R., Hauptfeld, R., Mellor, D. and Gallo, J. (2017) Leveraging the power of place in citizen science for effective conservation decision making. *Biological Conservation*, *208*, pp55–64.

Nielsen, J.V. and Jan Arvidsen. (2021) "Left to their own devices? A mixed methods study exploring the impacts of smartphone use on children's outdoor experiences". *International Journal of Environmental Research and Public Health* 18, no. 6: 3115. https://doi.org/10.3390/ijerph18063115

Omiyefa, M.O., Ajayi, A. and Adeyanju, L.O. (2015) Exploring ecopedagogy for the attainment of education for all in Nigeria. *Journal of Education and Practice*, 6(6) pp40–44.

Ouvry, A. and Furtado, A. (2020) *Exercising Muscles and Minds*, Second edition, London, Jessica Kinsley Publishers.

Parliamentary Office of Science and Technology (2014) *Environmental citizen science*. Parliamentary Office of Science and Technology (POST), Houses of Parliament, POST NOTE 476, August 2014. http://researchbriefings.parliament.uk/ResearchBriefing/Summary/POST-PN-476

Parsley, K.M. (2020) Plant awareness disparity: A case for renaming plant blindness. *Plants, People, Planet* 2: pp598–601. https://doi.org/10.1002/ppp3.10153

Powney, G.D., Carvell, C., Edwards, M., Morris, R.K.A., Roy, H.E., Woodcock, B.A., & Isaac, N.J.B. (2019) Widespread losses of pollinating insects in Britain. *Nature Communication*, *10*, p1018. https://doi.org/10.1038/s41467-019-08974-9

Price, C. A., & Lee, H. S. (2013) Changes in participants' scientific attitudes and epistemological beliefs during an astronomical citizen science project. *Journal of Research in Science Teaching*, 50(7) pp773–801.

Randall, R. and Brown, A. (2016) In time for tomorrow. *The Carbon Conversations Handbook*. https://graphicalert.com/wp-content/uploads/2014/07/klimaatgesprekken-werkboek-CC-BY-NC-ND-4.0.pdf

Richardson, M. and Sheffield, D. (2017) Three good things in nature: Noticing nearby nature brings sustained increases in connection with nature. *PsyEcology*, 8(1) pp1–32, DOI: 10.1080/21711976.2016.1267136

Rousell, D. and Cutter-Mackenzie-Knowles, A. (2020) A systematic review of climate change education: Giving children and young people a 'voice' and a 'hand' in redressing climate change, *Children's Geographies*, 18(2) pp191–208 DOI: 10.1080/14733285.2019. 1614532.

Sageidet, B.M., Almeida, C. and Dunkley, R. (2018) Children's access to urban gardens in Norway, India, and the United Kingdom. *International Journal of Environmental and Science Education*, 13(5), pp467–480.

Sanchez-Ortiz, K., Gonzalez, R.E., De Palma, A., Newbold, T., Hill, S.L., Tylianakis, J.M., Börger, L., Lysenko, I. and Purvis, A. (2019) Land-use and related pressures have reduced biotic integrity more on islands than on mainlands. bioRxiv, p. 576546.

Sandifer, P.A., Sutton-Grier, A.E. and Ward, B.P. (2015) Exploring connections among nature, biodiversity, ecosystem services, and human health and wellbeing: Opportunities to enhance health and biodiversity conservation. *Ecosystem services*, *12*, pp1–15.

Schneider, S.H. (1977) Climate change and the world predicament: A case study for inter-disciplinary research. *Climatic Change*, 1(1), pp21–43.

Smith, T.A. and Dunkley, R. (2018) Technology-nonhuman-child assemblages: reconceptualising rural childhood roaming. *Children's Geographies*, *16*(3) pp304–318.

Spence, A., Poortinga, W. and Pidgeon, N. (2012), The psychological distance of climate change. *Risk Analysis*, 32, pp. 957–972. https://doi.org/10.1111/j.1539-6924.2011.01695.x

Tsing, A.L. (2015) *The Mushroom at the End of the World: On the Possibility of Life in Capitalist Ruins.* Princeton, NJ: Princeton University Press.

Vanbergen, A.J. and Initiative, t. IP (2013), Threats to an ecosystem service: pressures on pollinators. *Frontiers in Ecology and the Environment,* 11, 251–259. https://doi.org/10.1890/120126

Wandersee, J.H. and Schussler, E.E. (1999) Preventing plant blindness. *The American Biology Teacher*, 61(2) pp82–91.

14

PAYING ATTENTION WITH MORE-THAN-HUMAN WORLDS

Field-visiting

Helen Clarke and Sharon Witt

Introduction

> Places tell stories
> Geography is a story
> (Fowler, 2017: p. 223)

'We must change the story; the story *must* change' (Haraway, 2016: p. 40).

We propose an expanded notion of geographical fieldwork that places relational thinking and understanding at the heart of geography education at all levels. Thinking and doing outdoor geographical education differently enhances fieldwork discourses and cultivates further ecological understandings. Massey (2005) contends, a 'place' is known through complex interrelations, so this chapter champions a geography education that attends to the interconnections of more-than-human and human worlds (Taylor, 2017). What happens when we invite a wider cast of participants into our understandings of geography fieldwork? We contemplate 'the agency, power and mystery of places' (Plumwood, 2008: p. 140) to propose a new conceptualisation of fieldwork.

We begin with a call for a paradigm shift from learning *about* to learning *with* the world. We then explore diverse influences on our thinking and practice to focus on relationality and materiality as integral to geographers' experiences with more-than-human fieldwork spaces. As we reimagine relationships with the world as situated, material, animate, and emergent, we share examples of encounters through poetic vignettes of educational assemblages (Pyyry, 2016). We rethink field skills as field responses and fieldwork as field-visiting – care-ful, response-able, polite, and wise. We present 'Pedagogies of Attention' and conclude by suggesting field-visiting offers something new for geographical education.

DOI: 10.4324/9781003248538-18

Being-with the world

A relational geography education seeks to disrupt contemporary Western notions of fieldwork, where children and teachers 'are placed in an inert and passive world waiting to be studied' (Puttick et al., 2018: p. 175). We embrace indigenous, feminist, posthumanist, and new materialist approaches, which situate children and teachers of geography as entangled *with* a sentient world. 'This requires a complete paradigm shift: from learning *about* the world to act upon it, to learning to *become-with* the world around us' (Common Worlds Collective, 2020: n.p.). Relational pedagogy positions learners as part of the world rather than separate from it. Relational fieldwork recognises the value in attending to more immediate, material, and place-based ways of experiencing the world and develops an ethics of care towards fieldwork spaces.

Our thinking is influenced by the work of academic geographers, such as Whatmore (2006) and Hinchliffe (2007), who include the more-than-human in their work. We are also inspired by non-Western thinking that complements, broadens, and deepens existing practices in geography education. We look to Bawaka et al. (2016) to understand place from indigenous perspectives and to grapple with the challenges of how wider ideas can inform our own. This synthesis of different ways of knowing is exemplified by Robin Wall Kimmerer (2013), an indigenous botanist and academic, who learns and teaches in the field with a spirit of humility, reciprocity, and gratitude. With place as teacher, and the more-than-human world as elders, wisdom is made visible through and activates connection with the world.

Relational fieldwork offers the potential 'to reimagine and relearn our place and agency in the world' (Common Worlds Collective, 2020). Teachers of geography can build on 'relations that many young children have with the world already' (Taylor, 2017: p. 12). This appreciative approach to children's more-than-human encounters is not an idealised, sentimentalised view of relationships with fieldwork spaces, rather it acknowledges the complexity of learning in a messy and unpredictable world. Relational fieldwork requires teachers to step back from setting agendas (Green and Somerville, 2015), to go with the flow, to follow the fluxes and movements of children's interactions with the world (Ingold, 2007). A less prescriptive approach is not without professional challenges such as constraints of curriculum, paucity of time, and low confidence and adds authenticity and spirit to children's fieldwork experiences.

Throughout the chapter, we illustrate our 'think-practice' (Thiele, 2014: p. 202) as relational entanglements in the form of poetic vignettes. Pacini-Ketchabaw et al. (2017: p. 1) position such entanglements as 'risky world encounters that affect us, provoke us to think and feel, attach us to the world and detach us from it, force us into action, demand from us, prompt us to care, concern us, bring us into question'. We use short, focused stories of place encounters that give a picture of field happenings at moments in time. These stories are generative and engage the reader by calling their own experiences and past learnings to mind. They also prompt new ways of thinking about place encounters and pedagogy. In seeking to be playfully diffractive with our ideas, the poetic vignettes draw from a range of field experiences, texts, images, theories, and questions to share more-than-human and human voices, traces, and presences. Diffraction is 'a way of understanding the world from within and as part of it' (Barad, 2007: p. 88). So, like diffractive waves, these ideas, concepts, and practices are brought into relation as part of an entangled state. This is 'a decolonising endeavor: to explore some of the implications of considering place as a partner in dialogue' (Bird Rose, 2002: p. 311).

Rethinking field spaces

The field is more than a physical space, rather it is 'a meeting place' (Cresswell, 2008) or 'a place of negotiation . . . between different elements' (Fors et al., 2013: p. 174). The weave of the field includes natural histories, atmospheres, weather, imaginings, archaeology, movement, affect, memories, folklore, traces, science, and intuition. To 'deep map' a place is to recognise this convergence of the more-than-human/human, as 'a simultaneity of stories-so-far' (Massey, 2005: p. 9). This requires a pedagogical response from teachers of geography that presents the 'world as multivocal, important, diverse and deserving of respect' (Blenkinsop et al., 2017: n.p.).

It is not our intention to position fieldwork-as-usual in binary opposition to relational fieldwork. We propose, in Table 14.1, a continuum, where fieldwork-as-usual might be expanded to embrace the more-than-human and include new ways of being, doing, and knowing fieldwork spaces. The left column identifies elements of fieldwork-as-usual practice. The right column suggests ways in which each practice might be made more relational. The emphasis of each fieldwork visit is then determined by the teacher of geography to enhance their curriculum making.

With these shiftings in mind, fieldwork becomes 'eventmental' (Page, 2020: p. 99), a 'place event' (Massey, 2005), which is situated, relational, material, animate, and in the moment. All these are interrelated.

Field spaces as situated

A perspective of the field as situated recognises the specificity and temporality of places. Within this framing, place is not a static collection of objects, rather it is a dynamic mix of activities, ideas, materials, things, forces, and intensities that come into relation with one another. Duhn (2012: p. 99) refers to 'place as assemblage', where ongoing movements of more-than-human gatherings constantly assemble, disassemble, and reassemble in fragmented and creative ways (Gannon, 2016: p. 132). Fieldwork sites change with different seasons, different times of day, with different weathers, as geographical assemblages are dynamic in the moment. Embodied encounters are situated in the 'here and now', and to recognise this 'bundle of trajectories' (Massey, 2005: p. 139) is to get to know a fieldwork space deeply. The following vignette offers children's collective responses inspired by a beach visit. Let's go to the shore, can you hear the waves?

TABLE 14.1 Moving from fieldwork-as-usual towards relational field spaces

Fieldwork-as-Usual	Continuum	Relational Fieldwork
Sites of knowledge extraction and skills practice	⟷	Sites of animation, engagement, and involvement (Witt, 2019)
Quest for knowledge	⟷	Making room for otherness and conceptual creativity (Blaise, 2016: p. 618)
Transmission of knowledge already known	⟷	Active inhabitation, experiencing with our bodies, and witnessing and evoking through practice (Mannion, 2020: p. 1369)
Places of human centrings	⟷	Places of posthumanist profusion (Taylor, 2016: p. 6)
Places of human knowing	⟷	Giving voice and recognising more-than-human elements within a site (Whatmore, 2006)
Pedagogical shifts from intrapersonal and interpersonal relationships	⟷	Intra-active relationships among all living organisms and the material environment (Lenz Taguchi, 2010: p. xiv)
Relations being forged in an already-given space	⟷	Relations as creative of spaces; they make spaces (Clarke and McPhie, 2014: p. 2002)
Known and usual ways reinforce unequal power relations	⟷	Undecidability shifts interactions onto 'a plane of equality' (Manson, 2001: p. 410)
Representation	⟷	Individuals learn to participate in a collective response-ability that is, a 'praxis of care and response' (Haraway, 2016: p. 105)

A fifteen-minute walk from school
Salterns Park near Hill Head
PO14 3QS
50.8137° N, 1.2230° W
A shingle beach
and promenade
South of Stubbington
West of Lee on Solent
South East of Titchfield
Hampshire
South Coast of England
Fawley Oil refinery amidst New Forest views to the right
At the head of Southampton Water

Gateway to the world
Overlooking the Solent
Home to major shipping lanes
and complex tidal patterns
Sheltered by the Isle of Wight.

The aforementioned beach encounter focuses on the specific detailed location of the place but clearly lies nested within much broader horizons. What connects this place to other places? What links materials in this place to other materials? What, and whose, stories emerge from possible trajectories within fieldwork spaces?

Field spaces as relational

A perspective of the field as relational acknowledges the collective agency and interdependence of all earthly beings, entities, and forces. For example, at the beach, children are entangled with tide; feet mark shoreline; pebbles are moved along sand; groynes edge spaces; sand disrupts drawings; shoes leave imprints; seaweed shelters crustaceans; air carries voices – processes are in action and an abundance of knowings emerge. Fieldwork places become lively and generative, as a pedagogical contact zone (Hodgkins, 2019), where children respond to and find new ways of relating and becoming-with the more-than-human world. This is an ongoing process of co-shaping and intermingling, which Haraway (2016: p. 13) refers to as 'worlding'. Worlding removes boundaries; it blends and enmeshes the participants in the fieldwork space. Let's return to the beach, where children's geopoetic encounters emerge:

We are Sea
we are entangled in Strandline
seaweed . . . pebbles . . . shells.
We are Sea
we are entangled with sand
as angels leaving traces of wing dances
waiting to disappear with incoming tide.
We are Sea
we are entangled with silence down on the shore
just the secret voice of waves that roll gently against shingle.
We are Sea
we are entangled with breeze whispering secrets
as sparkling sea glistens in sunlight.
We are Sea
we are entangled with reflective waters
as sea deposits memories.

At the shore, place and children meet, entangle, and 'intra-act' (Rautio, 2014). In intra-action place and children act together; experiences, ideas, and feelings emerge; changes and traces remain. All in the assemblage are transformed because of the encounter. Children come to know salty water, inter-tidal zones, marine processes, and landforms. What begins when we embrace such plurality, diversity, and complexity of experiences in fieldwork spaces?

The materiality of field spaces

Field spaces invite opportunities for direct material engagement. As educators, we tend to understand materials from a scientific, rational, or functional viewpoint and through predictable properties such as colour, shape, density, mass, friction, and gravity. Further, our understandings of materials are shaped by deeply rooted cultural dichotomies, including animate/inanimate and active/passive. These binaries lead us, often unconsciously, to think of ourselves as animate agents who act on passive, inanimate materials. This conception affects how we see materials, how we engage with them, and what we create with them.

Ingold (2020) calls for rigorous attentiveness to ongoing relations with lively materials. Materials have histories and stories and prompt events that allow children to ask questions and provoke enquiries. Materials of the field invite geographical possibilities. The following vignette diffracts children, beach, and theory to reveal material agencies:

> Children as curators
> explore, prospect, pay attention, notice.
> Thinking with driftwood,
> wood becomes a library, a meeting place, a boat to sail to unknown shores.
> Significance of objects-
> they have THING POWER!
> 'The curious ability of inanimate things to animate, to act,
> to produce effects, dramatic and subtle' (Bennett, 2010: p. 6).
> A collection,
> a gathering,
> making connections.
> In the company of
> a knotty cluster that is ever-changing,
> an on-going assemblage.
> Literal knowings and lyrical imaginings.
> Interesting stones, fossilised wood,
> 'Could this have been around at the time of the dinosaurs?'
> A shy shell hiding from the magical world of sealight,
> a fish band playing the soundtrack of the sea,
> 'Experimenting with ideas of matter as neither dull or static, but as alive and vibrant' (Youngblood Jackson and Mazzei, 2016: p. 96).

The field invites us to be open to a world of material experiences. Encounters that focus on process, not outcome, start from an understanding of places as agentic assemblages (Duhn, 2012), which show intensities and 'becomings' (Deleuze and Guattari, 1988). What if materials of the field shape us as much as we shape them?

The animacy of field spaces

A perspective of the field as animate acknowledges 'that the world is full of persons, only some of whom are human, and that life is always lived in relationship with others' (Harvey, 2006: p. xi). To be 'open to the world's aliveness, allowing oneself to be lured by curiosity, surprise and wonder' (Barad, 2012: p. 2) is to flourish and live life well in troubled times.

Merewether (2019) *thinks-with* 'enchanted animism' to (re)animate fieldwork spaces in times of climate change. 'An enchanted and lively world is one in which astonishment is part of everyday experience; a lively world which tells us what we care about, what is happening and what can be done' (Merewether, 2019: p. 247). The beach invites us again:

> *Children as followers . . . noticing, watching . . . listening . . . recording . . .*
> *What characters are found on the beach?*
> *What do the pebbles reveal?*
> *What stories will the pebbles tell?*
> *Routes of a seagull, a dog, a boat, a cloud . . .*
> *Tidal tales and strandline stories*
> *crossing thresholds*
> *in liminal spaces between*
> *sea and land.*

Animism opens the door to a world in which we can begin to negotiate life membership with an ecological community of kindred beings (Plumwood, 2009). Animate geographical field spaces offer possibilities of a sentient world, full of earthly entanglements. So, if we look differently 'what will no longer escape our notice?' (Kuby et al., 2019: p. 6).

Field spaces as emergent

A perspective of the field as emergent acknowledges that the field is contingent, fluid, and uncertain. The field cannot be taken for granted or predetermined, for spaces are messy, chaotic places of unknowing (Somerville, 2008). Knowing emerges with serendipitous encounters and requires a pedagogy of waiting to enable responses to the invitations of place. Fieldwork spaces generate ongoing emergent processes of 'becoming' between the more-than-human and geographers as relationships form, shift, and change. In this 'throwntogetherness' (Massey, 2005: p. 140) practices must be improvised in the moment; teachers have to be patient and trust that something generative will happen. The following vignette celebrates children's reflections on curios gathered as a shore museum. Let's make a final visit and enjoy beach revelations:

> *'You have to look and dig'*
> *'I learned that there are more colours than you think'*
> *'There aren't just pebbles beneath my feet*
> *but exciting things to explore'.*
> *'Oysters have little teeth at the back'*
> *'There's always things to be discovered'*
> *'The sun shines on the sea like a light bulb turning on'*
> *'We are all standing as one and we are never alone'.*

How can teachers of geography invite possibility and potentiality into fieldwork spaces? How can educators support children to respond with the field? How can geographers open themselves to what is yet to become?

Rethinking field skills as field responses

Field skills are an important element of geography education that help children to explore and explain places. Field skills bring children into relation. The following poetic vignettes follow the flow of a river studied from the classroom and then as a field visit:

> *A group of children locate River on a map of their locality*
> *they trace a thin blue line with their finger.*
> *They can see where River rises at its source and where it joins Sea.*
> *They can deduce River's direction of flow.*
> *They can research River's underlying geology.*
> *From photographs they know a little*
> *of what they might expect to see if they stood beside River.*
> *The children walk a section with River.*
> *Hands sketch through observation.*
> *Eyes identify and map River features – channel, bend, pool.*
> *Equipment helps humans see and measure.*
> *Tape allows a quantified width of River.*
> *Plastic duck races past marker points to reveal speed of Water.*
> *And . . . River splashes, River ripples, River holds other stories too.*

Field skills need to be taught, practised, and refined in places. Expertise emerges *with* place.

A list of constituent parts, of objective measurements, cannot fully indicate the complexity of relationships in an entanglement such as a river. So, we propose an expanded notion of field skills. One that includes both the traditional, more literal

TABLE 14.2 Developing expanded field responses

Field Skills	Expanded Field Responses
objective	**subjective**
literal	**lyrical**
simplistic	**complex**
surface	**deep**
separate	**relational**
generic	**situated**
data	**stories**
observing	noticing, attending, animating
measuring in standard units	qualitative scales, feelings and affects, values
comparing	appreciating similarity and difference, assimilating
classifying and naming	identifying, recognising, greeting
analysing	getting to know, searching, checking, revisiting,
interpreting	untangling, shedding light on, making sense of
questioning	asking, seeking,
recording	noting, jotting, breathing, collaging, journaling, tracing, curating, displaying
communicating	narrating, sharing, exchanging, communing, storying
naming	recognising, greeting
coming to know	interacting/intra-acting, relating, experiencing, connecting

collection of information about a place *and* simultaneously enables a range of more lyrical ways to respond with the field. Literal and lyrical dimensions complement and extend each other to offer deeper knowings of place. Table 14.2 proposes expanded ways of being, doing, and knowing fieldwork spaces.

This poetic vignette conveys *worlding-with* River and reveals rich stories of noticings and happenings:

The children have located River, mapped features of River,
and measured width and flow . . .
To get to know River, and to reveal River, other relationships emerge,
relationships that reveal 'riveriness':
Tom wonders at the pull of Water on his ankles as he stands in River's flow
Tilly stops to watch sparkles on Water's surface as sunshine emerges from clouds
above
Jon shrieks as cold Water seeps over his boots to wet socks and feet
Alice watches as Water flows into the tiny bottle she is holding,
and pours out again, as River once more.
Stick passes them, caught in a journey downstream.
Materiality, plants, critters, weather and moods are all bound up with children in the
field.
And children get to know River.

From viewing the world as full of objects to making kin with subjects is to honour difference, to respond to context, and to attend to the marginalised more-than-human world. Stengers (2014) retells the story of Gaia, not as a 'resource to be exploited or ward to be protected' but as a 'maker and destroyer' in her own right and not reducible to the sum of her parts (Haraway, 2016: p. 43). To walk with the field is to be a part of something bigger, to employ skills wisely, and to listen to all voices. Field skills become field responses. River's agency is revealed once more:

> *The children have located, mapped and measured River,*
> *they have also felt its wateriness, moved with River's processes.*
> *How might they tell of River? How might they voice River?*
> *With a story of noisy gulls chasing heron?*
> *Of a boat adrift in the channel?*
> *Of events witnessed and places still to come?*
> *With paint and paper?*
> *With poetic words and rhythms?*
> *As kin to be re-visited, re-encountered?*
> *Materiality, plants, critters, weather and moods are all bound up with children in the field.*
> *And River gets to know children.*

Rethinking fieldwork as 'field-visiting'

We propose a shift from fieldwork to 'field-visiting' as a new way to engage children in worldly geographical encounters. Field-visiting respects the 'intricacies' and 'richness' of the multiple ways children encounter places for they are 'in love with the world in a way that [is] energy – giving and interconnected' (Merewether, 2019. p. 234). So, what does it mean to 'visit'? To 'become part' of a place, at least for a while? To be in relation? To be connected? To be a part of geographical concepts, such as, place, space, scale, and human and physical processes? Field-visiting is a deepening of 'gaze' to enable different tellings that include more-than-human voices. It is a way to make the field visible and tell rich narratives of place.

Field-visiting is a different sort of practice – it is a 'curious practice' (Despret, 2005 in Haraway, 2016: p. 126), a thinking with other beings, animate and inanimate, and a propensity to find everything, especially the mundane and everyday, fascinating and thought provoking. Field-visiting is an entanglement, 'where everything is connected and affects everything else in a state of *one-ness*' (Lenz Taguchi, 2010: p. 39). Field-visiting is an 'ethico-onto-epistemological' practice – an appreciation of the intertwining of 'ethics, knowing, and being' (Barad, 2007: p. 185). We propose geographical field-visiting as storying a visit to a place and, like all good stories, field visits have a beginning, a middle, and an end. The processes of fieldwork storying are described in Table 14.3.

TABLE 14.3 Field-visiting as ethico-onto-epistemological practice: entering–dwelling–leaving

Ethics caring with the world	How do you enter into an encounter?	attentively – noticing and listening – with care and collective voice
Ontology being with the world	How do you dwell within the encounter?	attentively – how you are, how you travel, how you respond – slowly, politely, respectfully – with whimsy, imagination, reciprocity and humility
Epistemology knowing with the world	How are you transformed in relation to/by your encounter?	attentively – in multiple ways of knowing (embodied, emotional, material, lyrical) – with wisdom and gratitude

Care-ful, response-able field-visiting

How you enter into an encounter matters. Of course, expectations of safe behaviour and risk assessment procedures are essential practices for school field visits. In addition, we propose a nuanced emphasis on collective field ethics for '*response-ability*' that is, a 'praxis of care and response' (Haraway 2016: p. 105). Response-ability requires teachers of geography to model appreciative attention, to notice in respectful ways, and to recognise happenings in the field in the company of more-than-human kin. In '*Wild Pedagogies*' Jickling et al. (2018: p. 39) urge educators to give voice to entities and earthly processes and to 'foster practices and orientations that recognise voices of the more-than-human world and their educative significance'. Giving voice is to give witness to happenings in fieldwork spaces, encouraging children to be open and to accept invitations of place: to trace wave patterns; to immerse with weather events; to follow murmurations of birds in flight. To be poised to respond to the not-yet-known is challenging and demands 'joy, play, and response-ability to engage with unexpected others' (Haraway, 2015: p. 163). Let's join a care-ful field visit as children enter woodland:

> *A group of children arrive at Woodland in a state of anticipation.*
> *They don their 'cloak of silence' as they approach*
> *they enter calmly, singly and with reverence.*
> *They are known visitors as they have been here before,*
> *they offer greetings,*
> *they wait for familiar kin to appear*
> *they notice change, newness.*
> *They are open to what Woodland might reveal today:*
> *Bluebell coming into bud,*
> *pathways leading to margins,*
> *a group den as 'home' and curation spot,*
> *voices of Oak, Rook and Robin, Bramble and Woodcutter.*

Polite field-visiting

How you dwell with an encounter matters. Field-visiting positions geographers as 'wayfarers' (Ingold, 2007). Wayfaring requires us to tune into the rhythms of a sentient more-than-human world and respectfully give sustained attention (Ingold, 2007). Movement through place creates embodied, tacit ways of knowing that are forged along the way. Pausing with place opens time to *be-with* many beings, materials, and processes. To be a polite visitor is to recognise that 'landscape is loud with dialogues, with storylines that connect a place and its dwellers' (Spirn, 1998: p. 170). Polite geographers ask questions that are truly provoking. 'With good questions, even mistakes and misunderstandings can become interesting' (Haraway, 2016: p. 127).

Slow pedagogies enable sensory, embodied, material, and affective encounters over time: time for noticing and listening; time for exploring and thinking; time for making and being. Pausing and dwelling with spaces 'for more than a fleeting moment' (Payne and Wattchow, 2009: p. 16) cultivates a more-than-human geography community through attunement, proximity, openness, and receptivity. Slow pedagogies do not imply a lethargic or lazy attitude to more-than-human world exploration. In fact, embracing a deliberate and unhurried approach allows for opportunities to playfully engage, create, commune, share, build rapport, and recognise something of the specificity of field spaces.

Polite geographers do not rush, they show reciprocity with the more-than-human world, in 'ongoing interchange between body and the entities that surround it . . . a sort of silent conversation . . . in continuous dialogue' (Abram, 1997: p. 52). Wall Kimmerer (2013: p. 190) suggests 'we can respond through gratitude, through ceremony, through land stewardship, science, art and in everyday acts of practical reverence'. In these ways, field-visiting allows more-than-human participants and geographers to intra-actively shape what unfolds, even when field spaces are not who or what we expect them to be. For example, we have visited woodland to find bluebells, but it was beech trees that caught our attention; we paused with meadow expecting to see butterflies but were intrigued by grasshoppers. A polite field visitor is open to the surprise of the field and cultivates the capacity to respond. Let's spend more time with Woodland:

> *Children dwell with Woodland . . .*
> *they attend to vegetation*
> *construct biome models –*
> *canopy layer, herb layer, ground layer.*
> *There is no need to hurry, Woodland is patient*
> *Twists and tangles*
> *fallen branches, stumps.*
> *What happened here? Whose voice is that? What is next?*
> *Root-girl hesitates, pokes with stick, twists, turns, gives attention,*
> *brings fascination to material knowings.*
> *Clay Soil and sharp Flints greet fingers*

Chalk catches eyes.
Fresh shapes emerge
and new enquiries.

Wise field-visiting

How you leave an encounter matters – move thoughtfully and with gratitude. Stay attentive, as the 'practice of slowing down' opens the 'wisdom space within' (Hart, 2001: p. 26). Field-visiting embraces many ways of knowing, including more-than-curriculum knowings, embodied knowings, affective knowings, material knowings, and lyrical knowings. These knowings are inseparable; they are relational and generate expanded place wisdom. Field-visiting has the potential to be collectively transformational; for wisdom and for flourishing. Shifting fieldwork to field-visiting is a commitment to broaden *thinking-with* place that enriches the subject of geography, making it seemingly boundary-less (Angus et al., 2001). Working within disciplinary borderlands rejects the idea of a 'single linear narrative' of a place and reveals a world of multiple and contested meanings that provide 'a potential source of experimentation, creativity and possibility' (Giroux, 1991: p. 63). Field-visiting plays between the lines, in the nooks and crannies, and opens spaces that build different stories of relationship, care, and connection. Reimagining fieldwork as field-visiting 'opens the crack in the here and now' (Anderson, 2004: p. 705) and presents opportunities for deep encounters and bold geographical thinking. We prepare to leave Woodland:

> *Children engage in ceremonies of gratitude and reflection.*
> *Knowings are spoken aloud*
> *becomings are shared with Woodland cast*
> *and memories are treasured in story stones.*
> *Children leave Woods holding*
> *embodied encounters with Trees, Mud, Mist, Trails, and mystery –*
> *glimpse of Deer springing suddenly from undergrowth,*
> *drawing attention to ground cover,*
> *the discovery of a blackened mound, evidence of fire amongst green,*
> *Stump that became Den, a place of imaginings,*
> *Bluebells breaking steps and weaving paths,*
> *enigmatic shapes and possibilities . . . a dragon's nose?*
> *Children pass back through Boundary Gate, donning their worldly cloak.*
> *New knowings travel with them.*

Field-visiting and Pedagogies of Attention

Field-visiting has emerged from our diffractive readings of practice, conversations, and encounters. Field-visiting opens 'new pedagogical spaces of enchantment' through an ongoing engagement with the world, 'a development of embodied

TABLE 14.4 Pedagogies of Attention

Pedagogies of Attention
Modelling appreciative attention, noticing in a respectful way
Recognising & celebrating gifts in the company of more-than-human kin
Attending in generous reciprocity with the Earth
Being unhurried; pausing, lingering, dwelling
Being playful; embracing emergence, complexity and messiness
Being receptive to the invitations of a lively world
Being open to enchantment, serendipity and uncertainty
Being storymakers and storytellers; it matters what stories we tell
Attuning to relational encounters (*knowing-with*)
Noticing details of everyday places
Engaging actively and immersing all senses.

Source: Clarke and Witt, 2017.

skills of perception and action' and 'paying attention to experience as it is experienced' (Pyyry, 2016: p. 103). Attending to more-than-human worlds is essential and under-theorised. Field-visiting is all about attention: caring-with attention, being-with attention, knowing-with attention. It matters what we attend to. Field-visiting enacts behaviours of 'Pedagogies of Attention' (Clarke and Witt, 2017), which include literal dispositions of attunement and engagement and also offer lyrical ways of caring, being, and knowing with the world. Table 14.4 'Pedagogies of Attention' shows a manifesto of field practices, developed from research encounters with learners of all ages ranging from early years to adults.

Pedagogies of Attention bring geographers into relation with the world and open up possibilities for geographical learning rooted in ethical engagements, empathy, and care.

Field-visiting encompasses all the learning of fieldwork-as-usual. It also adds layers of complexity and nuance. Furthermore, such openness to serendipity courts surprise and mystery, as there are no guarantees as to precisely who, and what, might be encountered; nor to whom, and how, we might be called to respond.

Field-visiting offers something different

Field-visiting is a timely contribution to a different script of the field, where we 'turn up the colour and tune in to the world . . . with a positive energy and an attention to the exploration of alternate possibilities' (Geoghegan and Woodyer, 2014: p. 219). We have reimagined fieldwork and, in shifting towards field-visiting, we have illustrated entanglements of place–materials–children through poetic vignettes.

Contemporary times change what really matters. It is time to make space for new ways of acting, being, and thinking in geography education. Relational field-visiting opens more nuanced possibilities for geographical knowings. Be bold . . . be polite . . . go visiting . . . engage with relational narratives to embrace unfolding stories and to celebrate more-than-human profusion.

References

Abram, D. (1997) *The Spell of the Sensuous: Perception and Language in a More-Than-Human World*. New York: Vintage Books

Anderson, J. (2004) Talking whilst walking: A geographical archaeology of knowledge. *Area* 36 pp254–261

Angus, T. Cook, I. and Evans, J. (2001) A manifesto for cyborg pedagogy? *International Research in Geographical and Environmental Education* 10(2) pp195–201

Barad, K. (2007) *Meeting the Universe Halfway: Quantum Physics and the Entanglement of Matter and Meaning*. Durham, NC: Duke University Press

Barad, K. (2012) On Touching – The inhuman that therefore i am. *Differences* 23(3) pp206–223

Bawaka Country, Wright, S. Suchet-Pearson, S. and Lloyd Macquarie, K. Burarrwanga, L. Ganambarr, R. Ganambarr-Stubbs, M. Ganambarr, B. and Maymuru, D. Bawaka, and Sweeney, J. (2016) Co-becoming Bawaka: Towards a relational understanding of place/space. *Progress in Human Geography* 40(4) pp455–475

Bennett, J. (2010) *Vibrant matter: A political ecology of things*. Durham, NC: Duke University Press

Bird Rose, D. (2002) Dialogue with Place: Toward an Ecological Body, *Journal of Narrative Theory* 32(3) pp311–325

Blaise, M. (2016) Fabricated childhoods: uncanny encounters with the more-than-human. *Discourse: Studies in the Cultural Politics of Education* 37(5) pp617–626

Blenkinsop, S. Affifi, R. Piersol, L. and De Danann Sitka-Sage, M. (2017) Shut-Up and Listen: Implications and Possibilities of Albert Memmi's Characteristics of Colonization upon the 'Natural World'. *Studies in Philosophy and Education*, 36(3) pp349–365

Clarke, D.A.G. and Mcphie, J. (2014) Becoming animate in education: immanent materiality and outdoor learning for sustainability. *Journal of Adventure Education & Outdoor Learning*, 14 (3), 198–216

Clarke, H. and Witt, S. (2017) *A pedagogy of attention: a new signature pedagogy for educators*, British Educational Research Association Conference, University of Brighton, September 5–7th

Common Worlds Collective (2020) Learning to become with the world: Education for future survival, Education Research and Foresight Working Papers, UNESDOC library https://unesdoc.unesco.org/ark:/48223/pf0000374923

Cresswell, T. (2008) Place: encountering geography as philosophy. *Geography* 93(3) pp132–139

Deleuze, G. & Guattari, F. ([1988]/2013) *A Thousand Plateaus: Capitalism and Schizophrenia (B. Massumi, Trans.)*. London: Bloomsbury

Despret, V. (2005) Sheep do have opinions. In B. Latour and P. Weibel (Eds) *Making Things Public*, Cambridge, MA: MIT Press, pp360–368

Duhn, I. (2012) Places for pedagogies, pedagogies for places. *Contemporary Issues in Early Childhood* 13 (2) pp99–107

Fors, V., Å. Bäckström & S. Pink (2013) Multisensory emplaced learning: Resituating situated learning in a moving world. *Mind, Culture, and Activity* 20(2) pp170–183

Fowler, A. (2017) *Hidden Nature*. London: Hodder

Gannon, S. (2016) Local Girl Befriends Vicious Bear: Unleashing Educational Aspiration through a Pedagogy of Material- Semiotic Entanglement. In C.A. Taylor & C. Hughes (Eds) *Posthuman Research Practices in Education*. Basingstoke: Palgrave Macmillan, pp128–149

Geoghegan, H. & Woodyer, T. (2014) Cultural geography and enchantment; the affirmative constitution of geographical research. *Journal of Cultural Geography* 31(2) pp218–229

Giroux, H.A. (1991) Border Pedagogy and the politics of postmodernism. *Social Text*, 28, pp51–67

Green, M. and Somerville, M. (2015) *Children, Place and Sustainability*. Basingstoke: Palgrave Macmillan

Haraway, D. (2015) Anthropocene, capitalocene, plantationocene, chthulucene: Making kin. *Environmental Humanities* 6(1) pp159–165

Haraway, D. (2016) *Staying with the Trouble: Making Kin in the Chthulucene* (Experimental Futures), Durham, NC: Duke University Press

Hart, T. (2001) Teaching for wisdom. *Encounter: Education for Meaning and Social Justice*,14 (2), https://childspirit.org/wp-content/uploads/2011/09/Child-Spirit-Carrollton-GA-Teaching-for-Wisdom.pdf

Harvey, G. (2006) *Animism: Respecting the Living World*. New York: Columbia University Press

Hinchliffe, S.J. (2007) *Geographies of Nature*. London: Sage

Hodgkins, B. (Ed.) (2019) *Feminist Research for 21st Century Childhoods: Common Worlds Methods*. London: Bloomsbury

Ingold, T. (2007) *Lines: A Brief History*. London: Routledge

Ingold, T. (2020) *Correspondences*. Cambridge: Polity Press

Jickling, B. Blenkinsopp, S. Timmerman, N. & Sitka-Sage, M.D.D (Eds) (2018) *Wild Pedagogies Touchstones for Re-Negotiating Education and the Environment in the Anthropocene*. Cham: Palgrave Macmillan

Kimmerer RW (2013) *Braiding Sweetgrass: Indigenous Wisdom, Scientific Knowledge and the Teachings of Plants*. Minneapolis, MN: Milkweed

Kuby, C.R. Spector, K. &Thiel, J.J. (2019) *Posthumanism and Literacy Education. Knowing/Becoming/Doing Literacies*. London: Routledge

Lenz Taguchi, H. (2010) *Going Beyond the Theory/Practice Divide in Early Childhood Education*. Abingdon: Routledge

Mannion, G. (2020) Re-assembling environmental and sustainability education: orientations from New Materialism, *Environmental Education Research*, 26:9–10, pp1353–1372

Manson, S.M. (2001) Simplifying complexity; a review of complexity. *Geoforum*, 32, pp405–414

Massey, D. (2005) *For space*, London: Sage

Merewether, J. (2019) Listening with young children: Enchanted animism of trees, rocks, clouds (and other things)). *Pedagogy, Culture and Society*, 27(2), 233–250.

Pacini-Ketchabaw, V. Kind, S. & Kocher, L.L.M. (2017) *Encounters with Materials in Early Childhood Education*. Abingdon: Routledge

Page, T. (2020) *Placemaking: A New Materialist Theory of Pedagogy*. Edinburgh Published to Edinburgh Scholarship Online: May 2021 DOI: 10.3366/edinburgh/9781474428774.001.0001

Payne, P. G. & Wattchow, B. (2009) Phenomenological deconstruction, slow pedagogy, and the corporeal turn in wild environmental/outdoor education. *Canadian Journal of Environmental Education*,14, pp15–32

Plumwood, V. (2008) Shadow places and the politics of dwelling. *Australian Humanities Review*, 44 (2008): 1–9. http://australianhumanitiesreview.org/2008/03/01/shadow-places-and-the-politics-of-dwelling/[Accessed 23 February 2009]

Plumwood, V. (2009) Nature in the active voice. *Australian Humanities Review*, 46, http://australianhumanitiesreview.org/2009/05/01/nature-in-the-active-voice/ [accessed 24/08/18]

Puttick, S. Paramore, J. & Gee, N. (2018) A critical account of what 'geography' means to primary trainee teachers in England. *International Research in Geographical and Environmental Education*, 27(2) pp165–178

Pyyry, N. (2016) Learning with the city via enchantment: Photo-walks as creative encounters. *Discourse: Studies in the Cultural Politics of Education*, 37(1) pp102–115

Rautio, P. (2014) Mingling and imitating in producing spaces for knowing and being: Insights from a Finnish study of child-matter intra-action. *Childhood*, 21(4) pp461–474

Somerville, M. J. (2008) Waiting in the chaotic place of unknowing': Articulating postmodern emergence. *International Journal of Qualitative Studies in Education*, 21(3) pp209–220

Spirn, A. W. (1998) *The Language of Landscape*. London: Yale University Press

Stengers, I. (2014) Gaia, the Urgency to Think (and Feel) https://osmilnomesdegaia.files. wordpress.com/2014/11/isabelle-stengers.pdf [date accessed 21/08/21)

Taylor, A. (2017) Beyond Stewardship: Common world pedagogies for the anthropocene. *Environmental Education Research*, 23(10) pp1448–1461

Taylor, C.A. (2016) Edu-crafting A cacophonous ecology. In C.A. Taylor and C. Hughes (Eds) *Posthuman Research Practices in Education*, Basingstoke: Palgrave Macmillan, 5–24

Thiele, K. (2014) Ethos of Diffraction: New Paradigms for a (Post)humanist Ethics. *Parallax*, 20(3) pp202–216

Whatmore, S. (2006) Materialist returns: Practicing cultural geography in and for a more-than-human world. *Cultural Geographies*, 13(4) pp600–609

Witt, S. (2019) Becoming lost within relational, democratic geographical fieldwork spaces, Thesis for (EdD) in Education, University of Exeter, https://ore.exeter.ac.uk/repository/handle/10871/37268

Youngblood Jackson, A. and Mazzei, L. (2016) Thinking with an Agentic Assemblage in Posthuman Inquiry. In C. Taylor and C. Hughes (Eds) *Posthuman Research Practices in Education*, London: Palgrave Macmillan, pp93–107

CONCLUSION

15

MOVING FORWARDS

Strengthening engagement across the intersections between children, education, and geography

John H. McKendrick, Simon Catling, Mary Biddulph, and Lauren Hammond

Introduction

> . . . basically nothing is being done . . . despite all the beautiful words . . . .
> (Thunberg, 2019, 84)

We opened this book with the words of William Blake, a poet and artist, who lived and died in London (England) in the eighteenth and nineteenth centuries. When writing 'The School-Boy' as an adult in his early thirties, Blake encouraged us to critically consider the nature of childhood, and purposes and practices of education, in that time-space. It seems fitting to open our concluding chapter with the words of Greta Thunberg, a young person of Generation Z, who in her late teens has been at the forefront of a contemporary movement to raise awareness of the ecological and climate emergencies and press politicians and the general public to truly engage with them, which requires active consideration of the relationships between people and the Earth – as they are and as they might be.

Thunberg's comments highlight that although sometimes there has been fine sentiment which has articulated an aspiration to tackle the climate and ecological crises, there has also been somewhat limited action, almost amounting to deflection, by politicians, the media, and those in positions of power. As such, the fine sentiments of some can also be contrasted with those of others who have continued to (re)produce injustices to protect or further themselves and their interests (Latour, 2018).

As we write together from four locations in the UK – Glasgow, Oxford, Nottingham, and London – on a shared online document, we consider these issues of power, focus, and action and contend that the relationships between children, education, and geography are of critical importance to discussions about society, the Earth, and the future. Thus, our purpose in editing this book has been to rethink

DOI: 10.4324/9781003248538-20

the intersections between children, education, and geography and to outline the possibilities that result from this process.

As introduced in Chapter 1 (Hammond et al., 2022), several fields of geography have active and diverse interests in education. In this book, we focussed particularly, although not exclusively, on exploring the ways in which engagement across the fields of *children's and young people's geographies, geographies of education*, and *geography education* could be mutually enriching. In doing so, several interesting and important ideas emerged. In this conclusion, we open by exploring some of the ways in which ideas and perspectives can be enriched when the intersections across children, education, and geography are in focus. We progress to identify possibilities and potential for future work, first for the benefit of geographical education and then for using this geographical education to benefit the world beyond. Before concluding, we return to intersections to outline ways in which we could strengthen focus on them in future work.

Intersections

An intersection can be conceptualised as a space/point where ideas or things meet and connect. We found this term appealing in actively considering the relationships between fields which have often existed in degrees of separation from one another.

Education and educational institutions are so much a part of society and everyday life that sometimes they are taken for granted processes, places, and spaces (Brooks and Waters, 2017). However, it is this very nature of education – as being woven deeply into society – that tells us something about (our) human geographies (Ibid.). This is a theme that Waters and Brooks (2022) examine as they present the case for reconfiguring scholarship on international student mobility (ISM), arguing for coterminous exploration of interconnections that exist in the geographies of international student mobility across micro-, meso-, and macro-scales. The micro-spaces that Waters and Brooks seek to interconnect with other scales of analysis for ISM are the specific focus in Kraftl's (2022) examination of education spaces. Kraftl also opens his chapter with a cross-scalar connection – observing the wider relevance of the environment of an education space (micro-scale) to the UN Sustainable Development Goals (macro-scale focus) (Dodds et al., 2017).

The intersections which Waters and Brooks's (2022) chapter invite us to explore are not defined only within the subject area of ISM. Waters and Brooks also suggest that insights from the ISM of young adults might 'travel' to better understand the mobilities of younger children at the meso-scale, referring to the long-term social visitor pass in Singapore, and how it is used to attract families with young children from China. Their work enriches our understanding of geographies of education, including how institutions, cities, and (young) people's geographies shape and are shaped by, migration, education, and market forces. These geographies are significant to young people's everyday lives and experiences of the world. They are also potentially significant to those researching and teaching in (geographical) education, not least in supporting them in considering the educational backgrounds, social identities, and spatialities of their students and how and why the geographies of the young people they teach are important to teaching and learning (Hammond, 2022).

Here, we suggest that active consideration of geographies of education and education spaces (Kraftl, 2022) can support further examination of what Disney and Schleihe (2019: p. 195) describe as the 'intricate nature and complex lived reality at a nexus of care and control' that are (re)produced in, and through, institutions. These geographies are important to both young people's experiences of education and to educationalists making decisions about teaching and learning; for example, in considering how students are supported to progress in, and through, (geography) education and to support institutional and societal inclusion (Biddulph et al., 2022). Dunkley (2022) extends the notion of educational spaces beyond the physicality of buildings and considers ways in which school grounds, as an example of micro-geographic and micro-ecological spaces, can be utilised to enable children to experience 'intimate encounters' with nature. She reports the work of the 'spot-a-bee' project where young children's 'ecological consciousness' is developed through their engagement with the non-human world.

Actively considering the intersections between children, education, and geography has also reinforced the importance of considering the relationships between different phases of education. This is significant in supporting diverse academic and social transitions (GeogEd, 2021) and considering the affective and embodied experiences of educational spaces. This is a theme which has, for example, been powerfully explored by Awa Farah and Alice Aedy in the film *Somalinimo* (Guardian, 2020). As Hammond and Healy (2022) examine, truly engaging with students is of critical importance to these debates. Their account reflects the importance of teachers and teaching and the knowledge and skills a geographical education can develop and how this can be an affective experience for young people. However, their research also reveals a deep-seated frustration by students at public examinations and associated performativity pressures shaping teaching and learning in schools. Although the young people with whom they engage can be seen as a minority population (those who progress to a geography degree at university), they are an important subgroup. This is because these students are close to the frontiers of knowledge production in their university studies but also temporally close to their education in schools (Ibid.).

Possibilities and potential – reconfiguring geographical education through fusion

Biddulph et al. (2022) explain that a focus on knowledge and curriculum has been prominent in recent debate in geography education, considering what children and young people should learn in schools and why. Elsewhere in this book, Lambert and León (2022) argue the importance of teachers enabling students to engage with 'powerful' disciplinary knowledge in schools. Building on the work of Michael Young and others, Lambert and León conceptualise powerful knowledge as knowledge which has been created and tested in disciplines, which can take students beyond their everyday experiences and imaginations of the world. This knowledge may also allow students to view and engage with the everyday in new ways (Catling and Martin, 2011; Lambert and Biddulph, 2014; Roberts, 2017). This vision of a geographical education is one which centres the young person, with the educator challenged to enable them to move beyond the known and the comfort of the familiar, while working with it. McKendrick (2022) presents an alternative view of 'powerful knowledge', concluding by advocating the power and potential of young people engaging with knowledge in schools that focuses on the wider conditions that shape (their own) lives to support and empower them in confronting injustice. Ultimately, this means that teachers of geography require not only an understanding of the nature, philosophies, and practices of geography as a discipline but also an understanding of the lives and lifeworlds of those who they teach (Catling and Martin, 2011). Indeed, the potential of making greater use of children's everyday geographies to enrich geographical education and realise its potential has been a strong theme across many chapters in this book (for example, Kraftl, 2022; Catling and Pike, 2022).

In considering if, how, and why (geographical) knowledge may be considered 'powerful', it is also important to engage with the significant systemic injustices in who is producing knowledge and which/whose knowledge is deemed powerful and by whom (Dennis, 2021; Rudolph et al., 2018). This includes active consideration of in/justices related to language, spatial location, migration, 'race', gender, sexuality, class, age, and (dis)ability and how these intersectional identities and factors may shape who is able to gain access to 'the academy' or positions of power in geography and education. This in turn may shape the methods and ethics of knowledge production, the nature of the knowledge produced, how knowledge is shared, and disciplinary communities. Here, we contend that it is also important to consider the value and importance of studying subjects beyond engaging with knowledge. For example, as Hammond and Healy (2022) explore, studying geography can be an affective and embodied experience, which ultimately may change how a person acts in the world. In exploring these debates, it is also important to consider other purposes of schooling beyond learning subjects, including, but not limited to, play and socialisation and how they matter to children and society.

Although the relationships between education and the everyday lives and geographies of the children and young people who are taught have not been ignored (Roberts, 2014; Catling and Willy, 2018; Hammond and McKendrick, 2020), we argue that they have been under-explored in teacher education programmes (Catling, 2011, 2017) and that geographies of education are often being omitted from the knowledge debates.

The potential for research and work across children and young people's geographies, geographies of education, and geography education to enrich geographical education in schools has been evident throughout this book. Catling and Pike (2022), who acknowledge that children's geographies are personal, varied, and divergent, describe the importance of local experiences and what they call 'thoughtful geographies' of how children encounter and rationalise the world that they experience. Likewise, Clarke and Witt (2022) consider how through field-visiting children, rather than learning 'about' the world, can learn 'with' the world and come to understand themselves in relation to the more-than-human-world rather than apart from it. In these two chapters, the smorgasbord of children's geographies offers a rich array of possibilities for school geography, building on research in children's geographies. Similarly, in a wide-ranging review of the micro-spaces of schools, Kraftl (2022) first observes the performativity-focussed concern in education with how the school space is thought to impact on learning outcomes. However, he demonstrates how the micro-geographies of education spaces have relevance that extends far beyond this. First, he notes how societal values are communicated through design and then goes on to show how these can then be shaped and challenged by teachers and students through everyday and incidental acts. These power plays are not always deliberate, conscious, or explicit: it is the craft of the children's geographer or geographer of education, which makes them transparent. Against the prevailing view that schools are receptacles for educational projects conceived from afar (Morgan, 2019, 2022), we are challenged to view school environments as dynamic and never finished: shaped, but never simply determined, by the world beyond.

As with Biddulph et al. (2022) and Waters and Brooks (2022), Pirbhai-Illich and Martin (2022) implore the need to de/colonise geography. They situate their argument against the development of geographical education in Canada and the UK, before progressing to provide two concrete examples of how this can be achieved – drawing on 'funds of knowledge' and establishing critical spaces of belonging. Although focused on de/colonising, the same frameworks might also be usefully adapted to challenge other ways in which geography as a discipline has marginalised some students. This task should not be underestimated and we are reminded that challenges to what has prevailed must have a clear framework and focus, or there is a risk of merely decanonising, rather than (in their case) de/colonising the curriculum. As they ably demonstrate, achieving this sense of purpose is predicated on a critical analysis of how coloniality in geography has emerged.

Possibilities and potential – the world within for the world beyond

School education around the world is now in a major state of flux, with questions being raised about children's and young people's access to educational opportunities and experiences. These concerns are not new, as is evidenced by the ongoing global project of extending access to primary education (as part of the Sustainable Development Goals – United Nations, 2022). However, the challenges of sustaining school education during the COVID-19 pandemic heightened awareness of inequities in access and experience, even when provision purports to be universal. Rather than serve as a vehicle for social mobility through its promotion of the meritocratic ideal, education can be accused of exacerbating social inequalities

(Reay, 2018) and maintaining the prevailing social order. It need not be so. We have argued that school education has the potential to rethink its broader social purpose and to better consider the ways in which children and young people, and the discourses that shape their lives, can be drawn into the education debate.

The chapters in Section III of the book – *progressive geographies in education* – demonstrate how, in practical ways, more progressive views of school geography can be explored in schools and classrooms. Mitchell and Béneker (2022) demonstrate ways in which the intersection between school and academic geography enables teachers to think differently about what they teach to draw in the geographies of children and young people. Puttick et al. (2022) make a similar case arguing for a more proactive relationship between local universities' climate change research and schools to develop more place-specific and place-relevant information about climate change. Also in this section, Dunkley (2022) and Clarke and Witt (2022) in different ways argue for the development of more relational understanding in and through school geography – how learning 'things' geographically enables young people to see and value the place of the more-than-human world in their daily lives and vice versa. Both advocate pedagogies that draw in and on the lived experiences of young people. Relational pedagogies are also central to Pirbhai-Illich and Martin (2022) which they argue are more likely to develop young people's capacities to think critically about what they see, hear, and experience beyond the classroom. Such critical insights are, they contend, the route to enabling different ways of thinking about 'spaces, place and boundaries of geography'. (p. 160)

By focusing on subject matter (climate change, conceptions of 'home', microgeographies, and migration) and ways of doing (fictionalised stories, the GeoCapabilities approach, relational field-visiting, and collaborations through citizen science) each of these chapters in Section III exemplifies what has been a core argument throughout – that a progressive geographical education, while challenging to build and create, has the potential to be inclusive, meaningful, and empowering in enabling children to better understand injustices and to use geographical concepts to connect issues that touch upon or frame their lives, with wider issues with which humanity is grappling.

Our concern to consider the implications for geographical education for children and young people has been evident throughout. At the outset, Waters and Brooks (2022) observed separate realms for international students on the campus of Western higher education institutions. As they explain, this separation is both inadvertently facilitated by institutions and actively sought by some students. However, in a cross-scalar observation, they question whether this is consistent with the student experience that is being marketed by institutions. At a time when many Western higher education institutions are concerned with 'decolonising the curriculum', there is also a need to broaden focus and consider how the wider spaces and practices of the university must adapt to support a safe, just, and rich experience for all students, and one which does not reinforce perceived cultural differences while engaging with the value of diversity. How we teach is also pertinent to consider in this context: Kraftl (2022) uses the example of how material objects (a knotted piece of rope, rather than a ruler to facilitate measurement) were used in one Nigerian school to provide contextually appropriate learning. Similarly, both case studies in Biddulph et al. (2022) complement and reinforce the arguments of Waters and Brooks (2022), when it is shown how everyday practices in and around schools are not always conducive to enabling minority pupils to fit in (case study 1) and transitions to, and experience of, university do not always serve all students equally well (case study 2). As discussed earlier, Pirbhai-Illich and Martin (2022) offer a framework for de/colonialising our approach to geographical education.

McKendrick (2022) reworks a similar argument for economic disadvantage (which, of course, is also a greater risk for many children from minority populations across the globe). He implores us to use our geographical tools to better understand and tackle the wider injustices that confront children before they reach the classroom, challenging us that if we do not do so, we are 'part of the problem'. He suggests that realising the potential of geography might involve a keener focus on those who engage with geography before pupils choose geography as one of their specialisms in the later years of their studies. Furthermore, and as with Waters and Brooks (2022), there is also a need to rethink what, who, and how we teach – and with what effect – within our geographical education. The life experiences and social capital that learners bring when confronting geographies of poverty must inform our classroom practice. This may be discomforting where our adult constructions of place fail adequately to reflect the lived reality of children and young people. What is envisaged is a disruptively progressive school geography with a strengthened focus on the geography of injustice in the UK, which is both *for* and *with* hitherto disadvantaged pupils.

As Morgan (2022) demonstrates, not all these concerns are without precedent. Taking the 'long view' of the development of geographical education, he shows how a focus on children in school geography has taken on different forms at different times, reflecting the wider educational and societal contexts of the moment. Much of what we propose in this book has antecedents in *The Bulletin of Environmental Education* and its 'streetwork', the Young Children's Geography Project, and the Geography for the Young School Leaver project, which Morgan presents as evidence of children's late twentieth-century emergence as a presence in geographical education. However, the history is not a linear march towards 'progress'; rather, it is argued that a 'geography in contraflow' emerged in the early 1990s. Looking ahead, and consistent with what Waters and Brooks (2022) and McKendrick (2022) propose, Morgan (2022) describes the 'urgent task' of the geography teacher as developing a curriculum that provides pupils with a working analysis of the political and economic forces that are shaping and determining their lives, which raise questions not only about which geographical knowledge perspectives and content selections count but also about whose knowledge matters most and for what purpose(s).

This potential is not limited to tackling injustices, such as racist practices and poverty. Catling and Pike's (2022) call for a reproachment of children's geographies and primary school geography has the potential to strengthen place-making, belonging, and attachment to the neighbourhood. Given many children's restricted engagement with their neighbourhood, when children study their locality through geography, they not only contribute their own experience but can also extend and deepen their knowledge beyond it. Sensitively crafted by the teacher, this endeavour has the potential to foster their sense of identity with and belonging to 'their place', about which they come to recognise its diversity and the richness this offers about notions of place and of personal experience, alongside the intersections of children, adults (among them, their teachers and others in schools), and particular places.

These progressive geographies with children need not always focus on children. Although Biddulph et al. (2022) concur with others in this book that school geography should draw in children's and young people's geographies in a range of ways to better connect the school subject to the lives and experiences of the young people, their rationale for doing so is to allow other geographies – social, economic, cultural, and environmental – to fulfil their educative potential. Children and young people's geographies can be a means to an end, as well as being inherently valuable. They contribute to enabling (geographical) learning in all its variety, its challenges, and its benefits personally and societally.

Extending directions and intersections

The chapters in this book have set out and examined occurrences, meanings, and possible impacts of geographies at the intersections of children and young people, geography, and education. But there is much more to explore. Here we suggest just a few avenues which we feel will provide fertile and stimulating areas for investigation. Others could readily be added, and we hope that this book will motivate their study and pursuit by readers.

One of the key impacts in many children's lives is the effect of mobility. While Waters and Brooks (2022) have highlighted this for university students as young people, it is an aspect of many children's and young people's lives which requires and merits much fuller and deeper investigation. This is not only a matter for students in international schools, be they students from elsewhere in the world or home students sent to these schools. Parental decisions, perhaps influenced by companies and governments, can extend or limit not only the places in which children find themselves but also the opportunities for geography in their education and the notions of geography on which that is drawn. But mobility is a 'home grown' matter because children can find themselves moving from place to place, house to house, and school to school, for some very frequently. The everydayness of migration from one part of town to another, from urban to rural neighbourhoods or between different areas within a country may affect a sense of belonging and identity, as inter-country migration illuminates (Catling and Pike, 2022). Mitchell and Béneker (2022) reorientate the position of migration as part of the geography curriculum. Their chapter reports ways in which teaching young people about the process of 'home-making' can offer a more humanistic approach to understanding migration, connecting to young people's sense of belonging and identity construction. While exchanging place has long been of interest, the geographies of educational shifts and the nature and effect of the geography studied remain to be deeply explored, particularly with children. Geographical examinations of moving between schools and curricula and their physical and social environments and approaches to learning are largely an ignored dimension waiting to be addressed illuminatively.

In some parts of the world, but increasingly in those areas which thought themselves unaffected, environmental changes have heightened the need for preparedness by and contributions from children in dealing with damaging natural events and disasters, including pandemics and in war zones. Puttick et al. (2022) have identified the need to explore the intersections between local and global environmental issues and concerns, educational curricula, and the studies of investigative geographers, but we know little about how these dynamic interests interact and the effects they have on lives, learning, and places. This is an example of where the nature of the geography involved might be explored more rigorously within geography curriculum topics at all levels of education. While studying resources, such as clothing or fuels (and the links between them), there is a strong justice dimension that is less fully explored, to do with such matters as the impact of resource extraction and product manufacture on environments and people, the real nature and costs of providing goods from one place to another, and the impacts of this on people's lives and ecosystems' health (see www.followthethings.com/). These geographies of places near and far, systems of educational provision, and the lives of children and connected adults are significant intersections, which raise important issues of social and environmental justice concerning the ethics of living in the world and of making a living.

There are several serious questions to be pursued here about the intersections between who researches and teaches geography at all levels, why they are engaged in the subject, the geography that is being researched, promoted, and publicly examined within the subject's range of possibilities, the ways that this research might and can have an effect on curriculum at all levels, and how it may engage students, particularly those who are less interested in geography and taught by non-geographers. There is a connection here with what really matters to children and young people when it comes to their geographical studies, as well as for policymakers and school decision-makers. Children and young people's interest in and, for some, determined activism about the climate crisis and about sustainability and matters of social and environmental exploitation (Thunberg, 2019) has raised interest in which dimensions of the environment really matter, how educational institutions respond to these and frame them, and the levels at which children's and young people's voices are listened to, heard, and acted on.

While children's neighbourhood geographies have been widely explored and reported on, and there has been much advice to primary and secondary schools about teaching local geography, there remains a need to investigate what it is that children gain through the school-based exploration of their local area, in terms of their sense of geography, the nature of a particular geographical area, and what should and can be included in any geography curriculum. This is a key aspect of the exploration of intersectionality, which raises questions about teachers, other adults in school, and parents and the locality. It concerns the variety of experiences and perspectives they have and hold as well as ways in which they connect with the wider area, nationally, and the world. These dimensions are not independent of each other. For instance, children may go to a school because of the 'accident' of their home's place or as a result of parental decisions to move to a particular school catchment. Places are, in this day and age, interconnected globally. It is likely that children's and young people's geographies, the geography they study, and the school they attend are much wider but partial and impacted by access to and varied uses of, for instance, social media, and directed, opened up, or constrained in a variety of ways through their schooling, perceptions of place, and the valuing of geography. At a range of scales, the interplay of children, education, and geography will affect them differently and not necessarily consistently. These intersections require investigation.

Another potentially rich area of investigation is if, how, and why the geographies of educational institutions (often key spaces in children and young people's lives) shape, and are shaped, by children's geographies. Here, how educational spaces relate to the nature of (geographical) education that is constructed, and how education and educational spaces are experienced and perceived by children and young people are of interest to those working across the intersections between children, education, and geography. As Barker et al. (2010; p. 379) explain 'schools are but one of many specialised adult-constructed and controlled institutions that place children in contained zones and structure their space and time'. Whilst children and young people may seek to use educational spaces and shape them as their own, questions also exist as to how children can be best involved and empowered through policy, research, and practice in the design of educational spaces. Whilst there is interesting research in this area (den Besten et al., 2008), very often young people are conceptualised as 'citizens in waiting rather than citizens in their own right' (Starkey et al., 2014: p. 428), meaning they are not always fully engaged in discussions about education and educational debates by those in positions of power and influence. Further examinations of the relationships between geographies, rights, and education could lead to active consideration of how children could be socially, spatially, and legally empowered. Finally, as Kraftl (2014) has explored through critical examination of alternatives spaces of education (for example, forest schools and home schooling), geographical concepts, ideas, knowledge, values, and methods can support and enable examination of areas as varied as power relations in educational institutions and the relationships between children and nature in different educational spaces. Engaging with these debates is significant for educationalists, in considering how they best empower and engage children through pedagogy and support them in making progress through curricula design.

As schools become increasingly research -led, -engaged, and -active, and research is potentially conducted 'on, with, for, by' (Bodén, 2021: p. 1) children and young people, it is significant that ethics is considered as a fundamental underpinning of future work. Whilst education as a discipline has a rich body of work on ethics, institutions such as universities have ethical policies, and organisations such as the

British Educational Research Association (BERA, 2018) produce guidelines for ethical research, we also contend that consideration of geography and geographies is helpful to this endeavour. Active consideration of the identities and geographies of children, the spaces and places in which research is conducted, and the philosophies that underpin research with children can further support consideration of social and spatial inequalities, power relations, and intersectionality. As we work towards more just tomorrows, it is of critical importance children are seen and truly engaged.

These few examples illustrate the possibilities for developing the interrelationships between children, education, and geography. While Brock (2016) and others (von Benzon and Wilkinson, 2019; Jahnke et al., 2019; Freytag et al., 2022) hint at matters of learning, and Young et al. (2014) and Butt (2020) say little about the spaces of education, while offering a broad-based perspective on geography, the intersections between geography, children, and education have remained relatively unexplored until the chapters in this book. It is an arena which has much to offer, given that children, geography, and their education are intimately interconnected, whether it is admitted, even grudgingly, or not. It is important that it is pursued further and more deeply.

In conclusion . . .

Through rethinking the intersection between children, education, and geography, we have sought to enhance knowledge of how scholarship across children's geographies, geographies of education, and geography education can be mutually enriching. This will involve consideration of how children learn and make meaning through connecting with their geographies and how and why a progressive (geographical) education can support and empower young people in their lives and futures. Engaging with children's geographies and geographies of education has the potential to support schoolteachers and academics in becoming more informed about those they teach and how to truly engage with their experiences and perspectives in, and about, education. Work at these intersections can also help us to better understand the relationships between education, place, space, and young people.

This book has moreover suggested potential new directions for (geographical) education in schools, which recognise children and young people as social actors who shape, and are shaped by, the spaces and places they exist within. Our contention is ultimately that research and work across the intersection between geographies of education, children's geographies, and geography education can support and inform progressive futures for children and young people. Over to you.

References

Barker, J. Alldred, P. Watts, M. Dodman, H. (2010) 'Pupils or prisoners? Institutional geographies and internal exclusion in UK secondary schools'. *Area* 42(3) pp. 378–386.
Biddulph, M. Hopkins, P. Tate, S. (2022) 'Connecting children's and young people's geographies and geography education: Why this matters to and for children, education and

society' in Hammond, L. Biddulph, M. Catling, S. McKendrick, J. H. (eds.) *Children, education and geography: Rethinking intersections* (pp. 69–82). Abingdon: Routledge.

Bodén, L. (2021) 'On, to, with, for, by: Ethics and children in research'. *Children's Geographies*. https://doi.org/10.1080/14733285.2021.1891405

British Educational Research Association (BERA) 'Ethical Guidelines for Educational Research (fourth edition)' available at: www.bera.ac.uk/publication/ethical-guidelines-for-educational-research-2018 (accessed 01/04/2022).

Brock, C. (2016) *Geography of Education: Scale, space and location in the study of education*. London: Bloomsbury.

Brooks, R. Waters, J. (2017) *Materialities and mobilities in education* Abingdon: Routledge.

Butt, G. (2020) *Geography Education Research in the UK: Retrospect and Prospect – The UK case, within a global context*. Cham: Springer.

Catling, S. (2011) Children's geographies in primary school in Butt, G (ed.) *Geography, Education and the Future* (pp. 15–29). London: Continuum.

Catling, S. (2017) Not nearly enough geography! University provision for England's pre-service primary teachers. *Journal of Geography in Higher Education*, 43(3), pp. 435–458.

Catling, S., Martin, F. (2011) 'Contesting powerful knowledge: The primary geography curriculum as an articulation between academic and children's (ethno-)geographies'. *The Curriculum Journal* 22(3), pp. 317–36.

Catling, S., Pike, S. (2022) 'Becoming acquainted: Aspects of diversity in younger children's geographies' in Hammond, L. Biddulph, M. Catling, S. McKendrick, J. H. (eds.) *Children, education and geography: rethinking intersections* (pp. 83–101). Abingdon: Routledge.

Catling, S., Willy, T. (2018) *Understanding and teaching primary geography*. London: Sage.

Clarke, H., Witt, S. (2022) Field visiting: paying attention to a more-than-human world in Hammond, L. Biddulph, M. Catling, S. McKendrick, J. H. (eds.) *Children, education and geography: Rethinking intersections* (pp. 215–231). Abingdon: Routledge.

den Besten, O. Horton, J Kraftl, K. (2008) 'Pupil involvement in school (re)design: participation in policy and practice' *CoDesign*, 4(4), pp. 197–210, DOI: 10.1080/15710880802524946

Dennis, N. (2021) The stories we tell ourselves: History teaching, powerful knowledge and the importance of context in Chapman, A. (ed) *Knowing History in school: Powerful knowledge and the power of knowledge* (pp. 216–233). London: UCL Press.

Disney, T. Schleihe, A. (2019) 'Troubling institutions' in *Area* 51(2), pp. 194–199 https://doi.org/10.1111/area.12501

Dodds, F., Donoghue, D., Roesch, J. (2017) *Negotiating the sustainable development goals: A transformational agenda for an insecure world*. Abingdon; Routledge.

Dunkley, R. (2022) 'Looking closely for environmental learning: Citizen Science and Environmental Sustainability Education' in Hammond, L. Biddulph, M. Catling, S. McKendrick, J. H. (eds.) *Children, education and geography: rethinking intersections* (pp. 198–214). Abingdon: Routledge.

Freytag, T., Lauen, D., Robertson, S. (eds.) (2022) *Space, Place and Educational Settings*. Cham: Springer.

GeogEd [Geography and Education Research Group] (2021) 'Student transitions: Journeys into and through geography at university'. www.rgs.org/research/higher-education-resources/remarkable-(1)/ (accessed 01/04/2022).

Guardian (2020) 'Young, British and Somali at Cambridge University'. YouTube, 2 September. Accessed 26 February 2021. www.youtube.com/watch?v=GsPy1guOctw.

Hammond, L. (2022) 'Recognising and exploring children's geographies in school geography'. *Children's Geographies* 20(1), pp. 64–78.

Hammond, L. Biddulph, M. Catling, S. McKendrick, J. H. (2022) 'The child and their (geographical) education' in Hammond, L. Biddulph, M. Catling, S. McKendrick, J. H. (eds.) *Children, education and geography: Rethinking intersections* (pp. 3–8). Abingdon: Routledge.

Hammond, L. Healy, G. (2022) 'Student voice, democratic education and geography: Reflecting on the findings of a survey of undergraduate geography students' in Hammond, L. Biddulph, M. Catling, S. McKendrick, J. H. (eds.) *Children, education and geography: Rethinking intersections* (pp. 102–116). Abingdon: Routledge.

Hammond, L., McKendrick, J.H. (2020) 'Geography teacher educators' perspectives on the place of children's geographies in the classroom'. *Geography* 105(2), pp. 87–93.

Jahnke, H., Kramer, C., Meusburger, P. (eds.) (2019) *Geographies of schooling*. Cham: Springer.

Kraftl, P. (2014) 'What are alternative education spaces – and why do they matter?' *Geography* 99(3) pp128–138.

Kraftl, P. (2022) 'Geographies of educational spaces' in Hammond, L. Biddulph, M. Catling, S. McKendrick, J. H. (eds.) *Children, education and geography: Rethinking intersections* (pp. 36–48). Abingdon: Routledge.

Lambert, D. Biddulph, M. (2014) 'The dialogic space offered by curriculum- making in the process of learning to teach and the creation of a progressive knowledge-led curriculum'. *Asia-Pacific Journal of Teacher Education* 43(3), pp. 210–224.

Lambert, D. León, K. (2022) 'The value of geography to an individual's education' in Hammond, L. Biddulph, M. Catling, S. McKendrick, J. H. (eds.) *Children, education and geography: rethinking intersections* (pp. 117–133). Abingdon: Routledge.

Latour, B. (2018) *Down to earth: politics in the new climatic regime*. Cambridge: Polity Press.

McKendrick, J.H. (2022) Children's geographies and schools: Beyond the mandated curriculum in Hammond, L. Biddulph, M. Catling, S. McKendrick, J. H. (eds.) *Children, education and geography: Rethinking intersections* (pp. 49–66). Abingdon: Routledge.

Mitchell, D. Béneker, T. (2022) Teaching migration with a geographic capabilities approach – expanding children's concept of home in Hammond, L. Biddulph, M. Catling, S. McKendrick, J. H. (eds.) *Children, education and geography: Rethinking intersections* (pp. 182–197). Abingdon: Routledge.

Morgan, J. (2019) *Culture and the political economy of schooling: what's left for education?* London: Routledge.

Morgan, J. (2022) 'Where have all the cool places gone? Young people's geographies, schooling and the curriculum problem' in Hammond, L. Biddulph, M. Catling, S. McKendrick, J. H. (eds.) *Children, education and geography: Rethinking intersections* (pp. 134–148). Abingdon: Routledge.

Pirbhai-Illich, F. Martin, F. (2022) De/colonising the (geography) curriculum in Hammond, L. Biddulph, M. Catling, S. McKendrick, J. H. (eds.) *Children, education and geography: Rethinking intersections* (pp. 152–167). Abingdon: Routledge.

Puttick, S. Chandrachud, P. Chopra, R. Robson, J. Singh, S. Talks, I. (2022) 'Climate change education: following the information' in Hammond, L. Biddulph, M. Catling, S. McKendrick, J. H. (eds.) *Children, education and geography: Rethinking intersections* (pp. 168–181). Abingdon: Routledge.

Reay, D. (2018) Miseducation: Inequality, education and the working classes. *International Studies in Sociology of Education*, 27(4), pp. 453–456.

Roberts, M. (2014) Powerful knowledge and geographical education. *The Curriculum Journal*, 25(2), pp. 187–209.

Roberts, M. (2017) 'Geographical knowledge is powerful if . . .' *Teaching Geography* 42(1) pp. 6–9.

Rudolph, S. Sriprikash, A. Gerrard, J. (2018) Knowledge and racial violence: the shine and the shadow of 'powerful knowledge'. *Ethics and Education*, 13(1), pp. 22–38.

Starkey, H. Akar, B. Jerome, L. Osler, A. (2014) Power, pedagogy and participation: ethics and pragmatics in research with young people. *Research in Comparative and International Education*. 9(4), pp. 426–440.

Thunberg, G. (2019) *No one is too small to make a difference*. London: Penguin Random House.

United Nations (2022) *Sustainable Development Goals*. [online]. www.globalgoals.org/

von Benzon, N. Wilkinson, C. (2019) *Intersectionality and difference in childhood and youth*. Abingdon: Routledge.

Waters, J. Brooks, R. (2022) Geographies of education at macro-, meso-, and micro-scales: young people and international student mobility in Hammond, L. Biddulph, M. Catling, S. McKendrick, J. H. (eds.) *Children, education and geography: Rethinking intersections* (pp. 21–35). Abingdon: Routledge.

Young, M. Lambert, D. Roberts, C. Roberts, M. (2014) *Knowledge and the future school*. London: Bloomsbury.

INDEX

Note: Page numbers in *italics* indicate a figure and page numbers in **bold** indicate a table on the corresponding page.

For Product Safety Concerns and Information please contact our EU
representative GPSR@taylorandfrancis.com
Taylor & Francis Verlag GmbH, Kaufingerstraße 24, 80331 München, Germany